Processing Digital Images in Geographic Information Systems

A Tutorial Featuring ArcView® and ARC/INFO®

David L. Verbyla and Kang-tsung (Karl) Chang

**Processing Digital Images
in Geographic Information Systems**
A Tutorial Featuring ArcView® and ARC/INFO®
By David L. Verbyla and Kang-tsung (Karl) Chang

Published by:

OnWord Press

2530 Camino Entrada

Santa Fe, NM 87505-4835 USA

Copyright © Dave Verbyla and Kang-tsung (Karl) Chang

First Edition, 1997

SAN 694-0269

10 9 8 7 6 5 4 3 2 1

Printed in the United States of America

Library of Congress Cataloging-in-Publication Data

Verbyla, David L.
    Processing digital images in GIS : a tutorial featuring ArcView and ARC/INFO /
David L. Verbyla and Kang-Tsung Chang.
        p.   cm.
    Includes index
    ISBN 1-56690-135-9
    1. Geographic information systems. 2. ArcView. 3. ARC/INFO.
    I. Chang, Kang-Tsung. II. Title.
    G70.212.V47   1997
    910'.285--dc21   97-16685
                                                              CIP

Trademarks

ArcView GIS and ARC/INFO are registered trademarks of the Environmental Systems Research Institute (ESRI), Inc. Avenue and Spatial Analyst are ESRI trademarks. Additional ESRI trademarks appearing in this book include ARC/INFO GRID, ARCPLOT, ARCEDIT, ARC Macro Language (AML), ARCDOC, and ARCTOOLS. OnWord Press is a registered trademark of High Mountain Press, Inc. All other terms mentioned in this book that are known to be trademarks or service marks have been appropriately capitalized. OnWord Press cannot attest to the accuracy of this information. Use of a term in this book should not be regarded as affecting the validity of any trademark or service mark.

Warning and Disclaimer

This book is designed to provide information on processing digital images with special attention focused on processing images within the ArcView GIS Spatial Analyst extension and ARC/INFO. Every effort has been made to make the book as complete, accurate, and up to date as possible; however, no warranty or fitness is implied.

The information is provided on an "as is" basis. The authors and OnWord Press shall have neither liability nor responsibility to any person or entity with respect to any loss or damages in connection with or arising from the information contained in this book.

About the Authors

David L. Verbyla is an associate professor of remote sensing/GIS in the Department of Forest Sciences, University of Alaska-Fairbanks.

Kang-tsung (Karl) Chang is a professor of geography at the University of Idaho. He teaches cartography, GIS, and GIS macro programming.

Acknowledgments

The authors would like to extend a special thanks to technical reviewers Christian Harder and Randy Worch at ESRI for their helpful comments.

Thanks to the students at the University of Alaska and workshop participants at the Alaska Surveying and Mapping Conference for enduring draft versions of selected exercises incorporated in this book. Both the students and conference participants helped to improve both the content and presentation of the exercises.

We appreciate the support of the Bonanza Creek Long-term Ecological Research Site program which supplied the scanned aerial photographs and satellite imagery used in the exercises.

Finally, thanks to Barbara Kohl, acquisitions editor at High Mountain Press, both for accepting our book proposal for publication, and for her nitty gritty work as project editor.

OnWord Press...

Dan Raker, President
David Talbott, Acquisitions and Development Director
Dale Bennie, Information Publishing Group Director
Carol Leyba, Associate Publisher
Barbara Kohl, Acquisitions Editor
Daril Bentley, Project Editor
Jean Cooksey, Project Editor
Cynthia Welch, Production Manager
Michelle Mann, Production Editor
Lauri Hogan, Marketing Services Manager
Kristie Reilly, Assistant Editor
Lynne Egensteiner, Cover designer, Illustrator

Contents

Introduction   **1**

Typographical Conventions 2

Companion CD-ROM 3

   File Installation 5

   CD Files by Chapter..............................9

Part I. Image Display

Chapter 1

Grayscale and Color   **19**

Panchromatic Display 21

   Contrast Enhancements 22

Panchromatic Display Using ARC/INFO 24

Panchromatic Display Using ArcView 25

Grid Display 26

   Grid Display Using ARCPLOT 26

   Grid Display Using ArcView GIS Spatial Analyst 30

Color Display 36

   Color Models in ARC/INFO 36

   Color Image Display Using ARC/INFO 42

   Color Display Using ArcView 44

Image-Based Slide Shows 47

   Slide Show Using ARCPLOT 48

Hot Link in ArcView 50

Chapter 2

Surfaces   57

Perspective Views 58
Perspective Views Using ARCTOOLS 63
  Simple 3D View 63
  Hypsometric Coloring and 3D Views 71
  Draping Images and Coverages 75
Perspective Views Using ARCPLOT Commands 78
  Simple 3D Surface 79
  Draping Features on a Surface 80
  Hypsometric Coloring 81
Shaded Relief 82
Shaded Relief Display Using ARCTOOLS 82
Shaded Relief Display Using ARCPLOT 85
Shaded Relief Display in ArcView 86
  Variations in Shaded Relief Display Using ArcView . . 88

Part II. Image Data

Chapter 3

Scanning   93

Bits and Bytes 94
Color versus Grayscale Scanning 94
Binary Threshold Scanning 95
Scanning Density 97
  Computing Scan File Size Requirements 99
Packaging Scanned Images 99
  Generic Image Formats 99
  Proprietary Image Formats 100
Heads-up Digitizing Using ARCSCAN 101
Simple Raster to Vector Data Conversion 102
  Changing the Straightening Properties 106
  Changing Raster Line Parameters 107
  Arc Editing Environment 110
Using ARCSCAN Utilities with ARCEDIT 110

Chapter 4

Remote Sensing   113

Spectral Regions   114

Image Resolution   121

    Spatial Resolution   121

    Spectral Resolution   123

    Temporal Resolution   124

    Radiometric Resolution   124

Satellite Imagery   125

    Historic Data   125

    High Spatial Resolution Satellite Images   125

    Quality Control   126

Digital Orthophotographs   128

Displaying Remotely Sensed Images   128

    Single-Band   128

    Multi-Band Images   132

Part III Unwarping Images

Chapter 5

Map Projection and Coordinate Systems   139

Map Projections   139

    Commonly Used Map Projections   143

Datums   147

Coordinate Systems   148

    Universal Transverse Mercator   149

    Universal Polar Stereographic   150

    State Plane Coordinate   150

National Map Accuracy Standards   151

Map Projections and Coordinate Systems in ARC/INFO .. 151

Map Projections and Coordinate Systems in ArcView 153

Chapter 6

Image Rectification   157

Affine Transformation   158

Developing Affine Transformation Models:
A Simple Example 159
Image Adjustments By Affine Transformation 161
Unwarping Images in ARC/INFO 164
Transforming Scanned Maps in ARC/INFO 171
Rectifying Images 176
Resampling Options 176
Image Rectification Using ARC/INFO 179
Rectifying Grids Using ArcView 181

Part IV. Classifying Images

Chapter 7

Unsupervised Classification   187

Unsupervised Classification Using ArcView 188
Spectral Clustering 189
Assigning Cover Types to Spectral Classes 191
Unsupervised Classification Using ARC/INFO 194
Single Band versus Multiband Classification 194
Spectral Clustering Using ARC/INFO GRID 196
Mapping Spectral Similarity
with Hierarchical Cluster Analysis 196
Assigning Cover Types to Spectral Classes 199

Chapter 8

Supervised Classification   203

Training Area Selection 204
Using ARCEDIT to Delineate Training Areas 205
Developing Training Area Statistics 206
Maximum Likelihood Classification 208
Using ARC/INFO GRID 211

Chapter 9

Accuracy Assessment   215

Reference Data 216

Random Sampling of Individual Image Pixels 216
Using a GIS to Generate an Error Matrix 218
Stratified Random Sampling with ARC/INFO 218
Assigning Reference Data Values Using ARCEDIT. . . .222
Producing an Error Table with ARC/INFO 223
Assigning Reference Data ValuesUsing ArcView 224
Producing An Error Table
Using ArcView Spatial Analyst 227
Creating an Error Matrix from the Error Table 228
Accuracy Types from the Error Matrix 229

Part V. Grid Operations and System Integration

Chapter 10
Grid Management, Clipping, and Resampling 233

Management Operations 233
Describing, Copying, Renaming,
and Deleting Grids in ARC/INFO 234
Describing, Copying, Renaming,
and Deleting Grids in ArcView 235
Clipping and Stitching 236
Clipping in ARC/INFO GRID 236
Clipping with ArcView Spatial Analyst 237
Cutting Grids Using a Polygon Coverage
in ARC/INFO GRID 240
Stitching Grids in ARC/INFO GRID 240
Cutting and Stitching Grids
Using a Polygon in ArcView. 241
Resampling 241
Grid Resampling Using ARC/INFO 242
Grid Resampling Using ArcView 244

Chapter 11
Noise Removal with ARCSCAN Tools 247

Removing Speckles 248
Removing Noise Regions 252

ARCSCAN Sketch Tools 254
Interactive Image Classification 260

Chapter 12
Filtering and Eliminating Groups of Pixels   263
Grid Filtering 263
   Edge Enhancement 264
   Filters For Smoothing Grids 268
Eliminating Groups of Pixels 273
   REGIONGROUP and NIBBLE 273
   GRID 275
   Spatial Analyst 277

Chapter 13
System Integration   279
Automatic Batch Processing 280
   ARC/INFO Batch Jobs 280
   ARC/INFO GRID Batch Jobs 281
Communication Between ArcView and ARC/INFO..... 282

Index   289

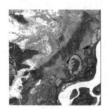

Introduction

The integration of diverse data sources in geographic information systems (GIS) is becoming increasingly common. GIS projects may encompass the use of traditional maps with point, line, and polygon map features, as well as grids, satellite images, digital elevation models (DEMs), digital orthophotos, and scanned data files. The latter "non-traditional" sources are generally referred to as raster data, or images and grids.

Examples of integrating images and grids with feature coverages are abundant in real world applications. Satellite imagery is an important data source for deriving land cover types, hydrography, and other surface features. Likewise, DEMs have provided the basis for dramatic 3D views and 3D draping. More recently, the use of digital orthophotos for updating feature coverages has become a routine operation in GIS maintenance. Aerial photographs and satellite images are also excellent tools for disaster monitoring and mapping. Disaster relief organizations and emergency planning agencies regularly use them in dealing with floods, fires, and oil spills. As high-resolution satellite images become more accessible to GIS users, applications of digital imagery are expanding into new areas such as mining, farm management, urban planning, and transportation.

The processing of images and grids in the Environmental Systems Research Institute's ARC/INFO and ArcView GIS differs from that of feature coverages. This book is a tutorial for GIS users who wish to quickly acquire proficiency in processing images and grids in GIS projects. Readers should be familiar with the basic functionality of ARC/INFO, ARC/INFO GRID, ARCPLOT, ARCEDIT, and/or Arc-View GIS and the Spatial Analyst extension to ArcView GIS. The book is written to ARC/INFO 7.x, and ArcView GIS 3.x.

Integration is taking place between GIS software packages. With the Spatial Analyst extension for version 3.0, ArcView users now have powerful tools for GIS analysis. With ArcTools, ARC/INFO users now have menu-based interfaces for editing, displaying, and analyzing spatial data. The trend toward integration is likely to continue and expand.

In your GIS work, you may choose to use ARC/INFO for some tasks and ArcView for others. For example, in ARC/INFO you can perform autotracing of scanned maps, supervised classification of satellite images, and multispectral analysis, but not in ArcView. On the other hand, the Spatial Analyst extension for ArcView GIS offers easy access to Avenue requests for grid data management and analysis. *Processing Digital Images* provides tutorials in how to perform selected tasks using ARC/INFO, or ArcView, or, in many cases, both. In addition, the book covers techniques for complicated tasks with the use of ARCTOOLS and ARCSCAN, both menu-driven user interfaces to ARC/INFO.

Typographical Conventions

The names of ArcView GIS and Spatial Analyst functionality interface items, such as menus, menu options, dialog box names, and dialog box options are capitalized.

> In the Add Theme dialog, Data Source Type must be set to Image in order to successfully select *spot-pan.bil*.

ARC/INFO commands (functions) and module names in running text are all capitalized.

GRID's ISOCLUSTER function is based on the iterative self-organizing data analysis algorithm (ISODATA). (Consult ARCDOC for details on the ISODATA clustering procedure.)

Command sequences in ArcView are often separated by a pipe (|).

With the grid theme *orth-grid* active, select Analysis | Map Calculator.

User input, and names for files, directories, variables, fields, themes, tables, coverages, expressions, and so on in running text are italicized.

These training area polygons have already been delineated in the *train-polys* coverage.

Select *tmrect.bil* as the image theme to add to the view.

Command lines entered into ARC/INFO and responses, AML code, ArcView Avenue code, and long expressions are shown in a monospaced typeface. The leading (space) between "word wrapped" lines is smaller than the leading between lines separated by hard returns.

```
Grid: gridcomposite rgb tmrect4 tmrect3 tmrect2 linear

/***Use text editor to create remap table for assigning shades to
each class.

Grid: &sys textedit colors.rmt
```

↝ **NOTE:** *Information on features and tasks that is not straightforward, immediately obvious or intuitive appears in notes.*

Companion CD-ROM

The companion CD contains four types of data: image files, ARC/INFO export files, documentation text files, and ASCII remap tables. Each data type is briefly described below.

The **ARC/INFO** *export files* are themes (coverages) and grids. Because coverages and grids vary slightly in format on different platforms (e.g., UNIX workstations versus Windows NT), you will use the platform-specific IMPORT utility that ships with ARC/INFO and ArcView GIS after the files are transferred to your hard disk.

Images can simply be copied from the CD to your workspace by using a system copy command or Windows File Manager. The image files are packaged as binary interleaved by line (BIL) or tag image file format (TIFF), and are accompanied by text files such as header, statistics, and world files. These files can simply be copied from the CD to your hard drive.

Documentation files can be copied from the CD to your hard drive. An ARC/INFO documentation file has been included for each chapter. Documentation files contain all ARC/INFO commands required to complete the ARC/INFO exercises in each chapter. Therefore, you can save time and avoid typos by opening an ARC/INFO window and a text edit window in which to load a documentation file. You can then copy every command from the text edit window and paste it to the ARC/INFO command window.

For example, to complete the ARC/INFO exercises for Chapter 1, you would open a text edit window and load the *display.doc* file into the window. Open another window in which to start ARC/INFO. At this point, you would copy the commands from the documentation file and paste them in the ARC/INFO window. Comment statements are preceded by the slash and asterisk (/***) symbols; characters following such symbols up to a hard return are ignored by the ARC/INFO command processor. Comment statements in documentation files describe the purpose or outcome of commands. An example follows.

```
arc /***Start ARC/INFO.

lg /***Get a listing of grids in the current workspace.
```

Remap tables can be copied from the CD to your hard disk. The remap tables are ASCII files used to assign grid shade colors and to reclassify grids based on the rules in the remap tables. In the documentation files, you will find references to using TEXTEDIT to edit these files. This step is optional because remap tables have been supplied on the CD; therefore, you need not create them. If you wish to change the assignments in the remap tables, you can use your system's text editor.

References to the TEXTEDIT command in documentation files means that you should use your system's text editor. For instance, the following command line indicates usage of your specific text editor to edit the *color.rmt* remap table for color assignments.

```
&sys textedit color.rmt
```

If you use the *vi* editor in UNIX, the command line would be changed to read as follows:

```
vi color.rmt
```

If you use the DOS editor, the same command would read as follows:

```
edit color.rmt
```

File Installation

We recommend that you store the files accessed and created throughout the exercises in a workspace or subdirectory on your hard drive. The following discussion of installing files from the companion CD to the hard disk is organized by operating system.

Windows 95 and Windows NT

1. Make a workspace or subdirectory called *workbook* on your hard drive, and then change directory to the new workspace. Both tasks can be accomplished in the File Manager or by using DOS commands.

2. Copy all files or selected files from the CD to the *workbook* workspace. Use the File Manager or the DOS copy command.

3. From DOS or a DOS window, navigate to the *workbook* directory and issue the following command to remove the read-only flag.

    ```
    ATTRIB    -R    /S
    ```

4. To import the ARC/INFO export files (files with *.e00* extensions) in **ArcView GIS**, use the *import71* program that ships with ArcView. Navigate to *import71.exe* in the *$AVHOME\bin* directory and double-click on it. (Alternatively, locate the Import71 icon in the Start menu, and double-click on it.)

    Note the type of file to be imported: cover, grid, or info. Refer to chapters and file names in the following tables to determine file type. Fill in the name of the file to be imported (including the *.e00* extension), and the name of the output file. The output file name is usually the same as the import file name minus the file extension.

5. To import the ARC/INFO export files (files with *.e00* extensions) in **ARC/INFO** under the Windows NT platform, you have two options. First, to import all ARC/INFO export files simultaneously, you can run the AML, *import-data.aml*, included on the companion CD. If you choose this option, it is assumed that you have copied all files from the CD to the *workbook* workspace. Input the following commands to run the AML.

    ```
    Arc /***Start ARC/INFO.

    Arc: &run import-data.aml
    ```

    The second option is to import the files individually. It is assumed that the selected files you wish to import have been copied to the *workbook* workspace. Examples of the three types of files to be imported appear below. Note that after entering each import command line you may wish to wait for the completion of the import operation.

```
arc /***Start ARC/INFO.

Arc: import cover <filename.e00> <filename>

Arc: import grid <filename.e00> <filename>

Arc: import info <filename.e00> <filename>
```

The command line for the three grid stack listings on the CD is slightly different. Command lines for importing the three stacks individually follow.

```
Arc: import info tm345stack.e00 tm345.stk

Arc: import info tm432stack.e00 tm432.stk

Arc: import info islands-stack.e00 islands.stk
```

↝ **NOTE:** *The import output files of four ARC/INFO export files on the CD are exceptions to the general rules for naming output files. The exceptions are emidalatlut.e00 (Chapter 2), tm345stack.e00 (Chapter 8), tm432stack.e00 (Chapters 8 and 9), and islands-stack.e00 (Chapter 10). Check the following tables by chapter for appropriate names.*

UNIX

1. Mount the CD-ROM on your system.

2. Make a workspace or directory called *workbook* on your hard drive. Change your working directory to *workbook*.

3. Transfer all files or selected files from the CD-ROM to the *workbook* directory.

4. To remove the read-only flag, input the following command line.

```
chmod  -R utw *
```

5. To import the ARC/INFO export files (files with *.e00* extensions) in **ARC/INFO**, you have two options. First, to import all ARC/INFO export files simultaneously, you can run the AML, *import-data.aml*, included on the companion CD. If you choose this option, it is assumed that you have copied all files from the CD to the *workbook* workspace. Input the following commands to run the AML.

```
arc /***Start ARC/INFO.

Arc: &run import-data.aml
```

The second option is to import the files individually. It is assumed that the selected files you wish to import have been copied to the *workbook* workspace. Examples of the three types of files to be imported appear below. Note that after entering each import command line you may wish to wait for the completion of the import operation.

```
arc /***Start ARC/INFO.

Arc: import cover <filename.e00> <filename>

Arc: import grid <filename.e00> <filename>

Arc: import info <filename.e00> <filename>
```

The command line for the three grid stack listings on the CD is slightly different. Command lines for importing the three stacks individually follow.

```
Arc: import info tm345stack.e00 tm345.stk

Arc: import info tm432stack.e00 tm432.stk

Arc: import info islands-stack.e00 islands.stk
```

6. To import the ARC/INFO export files (files with .*e00* extensions) in **ArcView GIS**, import the files before you load ArcView GIS. Use the *import71* program that ships with ArcView.

Note the type of file to be imported: cover, grid, or info. Refer to chapters and file names in the following tables to determine file type. Fill in the name of the file to be imported (including the .*e00* extension), and the name of the output file. The output file name is usually the same as the import file name minus the file extension. Example command lines follow.

```
$AVHOME/bin/import grid gridfilename.e00 gridfilename

$AVHOME/bin/import cover coverfilename.e00 coverfilename

$AVHOME/bin/import info infofilename.e00 infofilename
```

> **NOTE:** *The import output files of four ARC/INFO export files on the CD are exceptions to the general rules for naming output files. The exceptions are emidalatlut.e00 (Chapter 2), tm345stack.e00 (Chapter 8), tm432stack.e00 (Chapters 8 and 9), and islands-stack.e00 (Chapter 10). Check the following tables by chapter for appropriate names.*

CD Files by Chapter

Files stored on the companion CD are described below by chapter. You can copy all or selected files to a workspace on your computer's hard disk as you wish. Files used in exercises and referenced in particular chapters also appear at the beginning of respective chapters.

> **NOTE:** *All files with the .e00 extension are ARC/INFO export files. After you copy them to your workspace, they must be imported.*

Chapter 1. Grayscale and Color Display

File	Size (bytes)	Description
dogs.bil	90862	Panchromatic image of black and white
dogs.hdr	175	Header file describing *dogs.bil* image format.
dogs.stx	19	Statistics file for *dogs.bil* image.
coldgs.bil	1117440	Color image of dogs.
coldgs.hdr	176	Header file describing *coldgs.bil* image format.
coldgs.stx	56	Statistics file for *coldgs.bil* image.
color.bil	36	Simple image for color display exercise.
color.hdr	168	Header file describing *color.bil* image.
chena.bil	2325024	Scanned aerial photo of Chena River, Fairbanks, Alaska.
chena.hdr	178	Header file describing *chena.bil* image format.
chena.stx	57	Statistics file for *chena.bil* image.
compart44.tif	389977	Digital orthophoto of compartment 44, northern Idaho.
compart55.tif	304697	Digital orthophoto of compartment 55, northern Idaho.
slideshow.aml	445	ARC macro program for displaying four slides.
emida.rmt	587	Remap table for ARCPLOT grid shades.
emida2.rmt	468	Second remap table for ARCPLOT grid shades.
emida.cmf	267	Color map file for ARCPLOT grid shades.
compart.e00	13809	ARC/INFO polygon coverage of compartments, northern Idaho.
towns.e00	470480	ARC/INFO point coverage of towns in state of Alaska.
big-rivers.e00	1885575	ARC/INFO line coverage of major rivers in the state of Alaska.
dogs.e00	562391	ARC/INFO grid converted from *dogs.bil*.
emidalat.e00	350108	ARC/INFO grid of elevation values.
ak-shaded.e00	53768420	ARC/INFO grid shaded relief.
display.doc	5749	ARC/INFO documentation file for Chapter 1.

Chapter 2. Surface Display

File	Size (bytes)	Description
emidalat.e00	350108	Grid of elevation estimates for Emida, Idaho.
emidashade.e00	271924	Shaded relief grid of Emida, Idaho.
emidalatlut.e00	352	Info lookup table for grid shading. When importing export file, use *emidalut* as the output file name.

File	Size (bytes)	Description
emidacont.e00	165038	Line coverage of contours.
ak-shaded.e00	53768420	Elevation grid for the state of Alaska.
big-rivers.e00	1885575	Line coverage of major rivers in the state of Alaska.
towns.e00	470480	Point coverage of towns in the state of Alaska.
iditarod.e00	390382	Line coverage of the Iditarod trail in the state of Alaska.
surfaces.doc	2159	ARC/INFO documentation file for Chapter 2.

Chapter 3. Scanning

File	Size (bytes)	Description
hoytmtn.tif	8249475	TIFF scan of Hoyt Mountain, Idaho, 7.5-minute quad.
hoytmtn.tfw	222	World file for Hoyt Mountain, Idaho, scan.
hoytmtn-tics.e00	1347	UTM tic coverage for Hoyt Mountain, Idaho, quad.
scanning.doc	1114	ARC/INFO documentation file for Chapter 3.

Chapter 4. Remote Sensing

File	Size (bytes)	Description
hispat.tif	90445	High spatial resolution image of Chena River, Fairbanks, Alaska.
medspat.tif	90445	Medium spatial resolution image of Chena River, Fairbanks, Alaska.
lowspat.tif	90445	Low spatial resolution image of Chena River, Fairbanks, Alaska.
ortho.bil	1678644	Digital orthophoto of northern Idaho.
ortho.blw	222	World file for digital orthophoto of northern Idaho.
ortho.hdr	179	Header file describing digital orthophoto of northern Idaho.
ortho.stx	19	Statistics file for digital orthophoto of northern Idaho.
tmrect.bil	1191330	Landsat Thematic Mapper image of Alaskan interior.
tmrect.blw	222	World file for Landsat Thematic Mapper image of Alaskan interior.
tmrect.hdr	176	Header file describing Landsat Thematic Mapper image of Alaskan interior.
tmrect.stx	90	Statistics file for Landsat Thematic Mapper image of Alaskan interior.
veg-polys.e00	94607	Vegetation polygon coverage for digital orthophoto of northern Idaho.
remote-sensing.doc	2044	ARC/INFO documentation file for Chapter 4.

Chapter 5. Map Projection and Coordinate Systems

File	Size (bytes)	Description
usa.e00	3759048	ARC/INFO line coverage of United States.
map-project.doc	1382	ARC/INFO documentation file for Chapter 5.

Chapter 6. Image Rectification

File	Size (bytes)	Description
npole.tif	1049365	TIFF image of scanned North Pole quad.
spot-pan.bil	1913120	SPOT HRV Panchromatic satellite image.
spot-pan.hdr	179	Header file describing SPOT HRV image.
spot-pan.stx	18	Statistics file for SPOT HRV image.
scanquad.bil	582628	BIL image of scanned Fairbanks, Alaska, D2 quad.
scanquad.hdr	177	Header file describing Fairbanks, Alaska, D2 quad image.
scanquad.stx	14	Statistics file for Fairbanks, Alaska, D2 quad image.
rectifyscript.txt	1478	ArcView Avenue script for rectification.
gps-rds.e00	64793	Line coverage of logging roads.
longlat-tics.e00	1435	Tic coverage in geographic coordinates.
utm-tics.e00	1660	Tic coverage in UTM coordinates.
rectification.doc	3060	ARC/INFO documentation file for Chapter 6.

NOTE: The *spot-pan.bil* file is reproduced on the CD with the permission of SPOT IMAGE Corporation. © CNES/SPOT IMAGE.

Chapter 7. Unsupervised Classification

File	Size (bytes)	Description
islands.bil	956664	BIL image of scanned, rectified color infrared photo.
islands.blw	222	World file for *islands.bil* image.
islands.hdr	176	Header file describing *islands.bil* image.
islands.stx	60	Statistics file for *islands.bil* image.
yellow.rmt	98	Remap table for assigning yellow color with GRIDQUERY command.
reclass.rmt	248	Remap table for reclassifying spectral classes to land cover classes.
cover-types.rmt	199	Remap table for assigning grid shades to land cover type classes.

File	Size (bytes)	Description
islandsc1.e00	2044057	Grid corresponding to channel 1 or band 1 of scanned aerial photo.
islandsc2.e00	2057692	Grid corresponding to channel 2 or band 2 of scanned aerial photo.
islandsc3.e00	1995844	Grid corresponding to channel 3 or band 3 of scanned aerial photo.
spectral-map.out	3949	Dendrogram output based on 25 spectral classes.
unsupervised.doc	3763	ARC/INFO documentation file for Chapter 7.

Chapter 8. Supervised Classification

File	Size (bytes)	Description
trn-grid.rmt	460	Remap table for color assignments to training areas.
superv.rmt	386	Remap table for reclassifying training classes to land cover classes.
colors.rmt	350	Remap table for assigning grid shades for land cover type classes.
train-polys.e00	10655	Training areas polygon coverage.
tmrect2.e00	723988	Landsat Thematic Mapper band 2 grid.
tmrect3.e00	723616	Landsat Thematic Mapper band 3 grid.
tmrect4.e00	724692	Landsat Thematic Mapper band 4 grid.
tmrect5.e00	723743	Landsat Thematic Mapper band 5 grid.
tm345stack.e00	352	Stack listing of themes created in GRID. When importing ARC/INFO export file, use *tm345.stk* as output file name.
tm432stack.e00	352	Stack listing of themes created in GRID. When importing ARC/INFO export file, use *tm432.stk* as output file name.
supervised.doc	3317	ARC/INFO documentation file for Chapter 8.

Chapter 9. Classification Accuracy Assessment

File	Size (bytes)	Description
ranpts.e00	49680	Point coverage of randomly located sample points for reference data.
ref-grid.e00	1362237	Grid of reference data "truth."
class-grid.e00	602943	Classified grid "predictions."
tm432stack.e00	352	Stack listing of themes created in GRID. When importing ARC/INFO export file, use *tm432.stk* as output file name.
tmrect.bil	1191330	Landsat Thematic Mapper satellite image.
accuracy.doc	2094	ARC/INFO documentation file for Chapter 9.

Chapter 10. Grid Management, Clipping, and Resampling

File	Size (bytes)	Description
ak-dem.e00	66970279	Digital elevation grid for most of Alaska.
small-island.e00	5391	Polygon coverage of a small island.
islandsc1.e00	2044057	Grid corresponding to channel 1 or band 1 of scanned aerial photo.
islandsc2.e00	2057692	Grid corresponding to channel 2 or band 2 of scanned aerial photo.
islandsc3.e00	1995844	Grid corresponding to channel 3 or band 3 of scanned aerial photo.
islands-stack.e00	358	Stack listing of themes created in GRID. When importing ARC/INFO export file, use *islands.stk* as the output file name.
grid-mgmt.doc	2822	ARC/INFO documentation file for Chapter 10.

Chapter 11. Noise Removal with ARCSCAN Tools

File	Size (bytes)	Description
npole.e00	28401463	Grid from scanned topographic map of North Pole, Alaska, quad.
npole.tif	1049365	Scanned topographic map for the North Pole, Alaska, quad.
class-grid.e00	602943	Classified grid of land cover to be corrected with GRIDEDIT.
class.rmt	349	Remap table for assigning colors to land cover exercises.
gridedit.doc	1713	ARC/INFO documentation file for GRIDEDIT portion of Chapter 11.

Chapter 12. Filtering and Eliminating Groups of Pixels

File	Size (bytes)	Description
ortho.bil	1678644	Digital orthophoto for edge enhancement.
ortho.blw	222	World file for digital orthophoto.
ortho.hdr	179	Header file describing digital orthophoto.
ortho.stx	19	Statistics file for digital orthophoto.
class-grid.e00	602943	Classified grid of land cover to be filtered or smoothed.
spot-pan.bil	1913120	SPOT Image HRV satellite image.
filter.doc	1501	ARC/INFO documentation file for Chapter 11.

NOTE: The *spot-pan.bil* file is reproduced on the CD with the permission of SPOT IMAGE Corporation. © CNES/SPOT IMAGE.

Chapter 13. System Integration

File	Size (bytes)	Description
scanquad.bil	582628	BIL image of scanned Fairbanks, Alaska, D2 quad.
scanquad-world	222	File to be renamed to *scanquad.blw* world file.
scanquad.hdr	177	Header file describing Fairbanks, Alaska, D2 quad image.
scanquad.stx	14	Statistics file for Fairbanks, Alaska, D2 quad image.
fillsink.txt	3924	ArcView Avenue script for filling with ARC/INFO GRID FILL.
emidalat.e00	350108	Grid of elevation estimates from Emida, Idaho.
tmrect3.e00	723616	Landsat Thematic Mapper band 3 grid.
tmrect4.e00	724692	Landsat Thematic Mapper band 4 grid.
tmrect5.e00	723743	Landsat Thematic Mapper band 5 grid.
tm345stack.e00	352	Stack listing of themes created in GRID. When importing ARC/INFO export file, use *tm345.stk* as output file name.
batch.doc	836	ARC/INFO documentation file for Chapter 13.

Part 1
Image Display

1

Grayscale and Color

A digital image consists of a grid of cells. Each grid cell is called a *pixel* (picture element), and every pixel contains a number or numbers ranging from 0 to 255. A digital image is like a photograph. If you were to examine a photograph under a microscope, you could see grains from the film emulsion, but the image would not be recognizable. The image is decipherable only when thousands of film grains are visible. Pixels are analogous to film grains. Because only a few hundred pixels are displayed in the following digital image, it is virtually impossible to recognize the objects in the image.

Sample digital image.

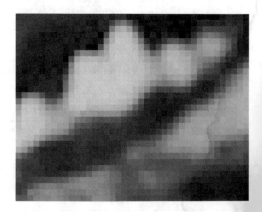

When thousands more pixels are displayed, the image becomes recognizable.

Same digital image displayed using thousands of pixels.

In this chapter grayscale and color images are displayed. The files from the companion CD listed below will be used in exercises and as illustrations.

dogs.bil	Grayscale image depicting two dogs.
dogs.hdr	Header file describing *dogs.bil* image format.
dogs.stx	Statistics file for *dogs.bil* image.
coldgs.bil	Color image depicting two dogs.
coldgs.hdr	Header file describing *coldgs.bil* image format.
coldgs.stx	Statistics file for *coldgs.bil* image.
color.bil	Color image.
color.hdr	Header file describing *color.bil* image.
chena.bil	Scanned color aerial photograph of the Chena River area in Alaska.
chena.hdr	Header file describing *chena.bil* image format.
chena.stx	Statistics file for *chena.bil* image.
compart44.tif	Digital orthophoto of compartment 44 (forestry management unit) in Chena River area, Alaska.
compart55.tif	Digital orthophoto of compartment 55 (forestry management unit) in Chena River area, Alaska.

slideshow.aml	Macro to display slide show created in ARC/INFO.
compart	Coverage containing forestry management units.
towns	Point coverage of cities, towns, and villages in Alaska.
big-rivers	Line coverage of major rivers in Alaska.
dogs	Grid based on dogs.bil.
emidalat	Grid of elevation values near Emida, northern Idaho.
ak-shaded	Shaded relief grid of most of Alaska produced using the ARC HILLSHADE command.
emida2.rmt	Remap table for Emida area, northern Idaho.
emida.cmf	Color map file for Emida area, northern Idaho.
emida.rmt	Remap table for ARCPLOT grid shades.
display.doc	ARC/INFO documentation file.

Panchromatic Display

A *grayscale* or panchromatic image is displayed by using pixel values to control screen brightness, also known as video intensity. For example, a pixel containing a zero value will appear black on screen, while a pixel containing a value of 255 will be bright white on screen. The lower the pixel value, the darker the gray. A histogram of pixel values from the previous image of a black dog appears below.

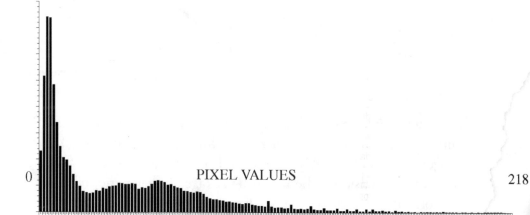

Histogram of pixel values ranging from zero to 218.

Note that if you use pixel values to directly control screen pixel intensities, the intensity range is zero to 218. This type of display is sometimes called *identity display*. Panchromatic display is improved by *contrast enhancement*, which applies to the full range of video intensity (0 to 255) to an image.

Contrast Enhancements

Linear Stretch

A common approach to contrast enhancement is the use of a linear contrast stretch. A linear stretch maps the minimum pixel value in an image to black (screen intensity of 0), and the maximum pixel value to bright white (screen intensity of 255). The following figure illustrates this method.

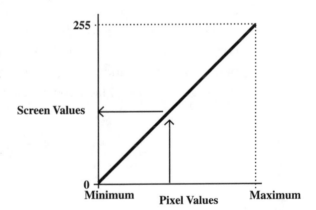

Min/max linear contrast stretch function.

Geographic information systems (GIS) software typically reads the minimum and maximum values of an image, which are stored in an image statistics file. In ARC/INFO and ArcView, this statistics file has an *.stx* file extension if the image is packaged in a generic *.bil*, *.bsq*, or *.bip* format. The screen intensity for any pixel value can be computed using the following formula:

```
screen intensity = [(pixel value - minimum value) / (max - min)] * 255
```

However, image contrast will not be markedly improved when the minimum and maximum represent rare pixel values. For example, consider an image almost entirely comprised of pixel values ranging from 100 to 150 except for a cloud with pixel values of 245 and a canyon shadow with pixel values of 5. Applying the min/max linear stretch in this case would not greatly improve the image contrast. Fortunately, there is a simple solution to this problem.

If the mean and standard deviation of the pixel values are stored in an image's statistics file, then a GIS will typically use these values to establish an upper and lower limit for contrast stretching. These limits are confidence limits based on statistical probability theory. Assuming that the distribution of pixel values is bell shaped, approximately 95 percent of all image pixels will have values within two standard deviations of the mean image pixel value. Thus, the upper and lower limits are typically calculated in ARC/INFO and ArcView as summarized below.

- upper value = mean + (2 * standard deviation)
- lower value = mean - (2 * standard deviation)
- screen intensity = [(pixel value - lower value) / (upper value - lower value)] * 255

In the case of the previous image of a black dog, the mean pixel value is 76.6, and the standard deviation, 23.1. Suppose that the face of the dog is made up of pixels with values of 62 and 66. If you display the image using the identity option, the contrast between these pixel values will be only four grayscale units. The image contrast can be enhanced by applying a 95 percent linear contrast stretch using the previous formulas.

ARC/INFO and ArcView use the mean and standard deviation in the image statistics file (*filename.stx*) to automatically perform a linear contrast stretch. ArcView also allows you to graphically control the contrast stretch through the Legend Editor.

Histogram Equalization

Under the equal area or histogram equalization method of contrast enhancement, each screen video intensity of 0 through 255 is assigned to approximately the same number of image pixels. Histogram equalization provides high contrast to the image pixel values that occur most often. For example, the following illustration shows the result of applying histogram equalization contrast to the previous image of two dogs. The contrast is enhanced because many of the image pixel values are in the high (white dog) and low (black dog) ranges.

Histogram equalization contrast applied to the dogs grid.

Panchromatic Display Using ARC/INFO

You can display images in ARCPLOT or ARCEDIT by using the IMAGE command. An example follows.

```
Arcplot: display 9999 4 /***Full-screen canvas.

Arcplot: mapextent image dogs.bil /***Scale canvas to fit the image.

Arcplot: image dogs.bil /***Display the image to the canvas.
```

ARC/INFO will display the .*bil* image using a contrast stretch if the image has a companion statistics file. The statistics file should have an .*stx* extension. For instance, examine the text file called *dogs.stx*. ARC/INFO expects the statistics file to contain the following information: band number, minimum, maximum, mean and standard deviation. If ARC/INFO finds the mean and standard deviation values, it will perform a 95 percent linear contrast stretch when you issue the IMAGE command. If ARC/INFO finds only the minimum and maximum values, then it applies the min/max linear contrast stretch when you issue the IMAGE command.

ARC/INFO also allows histogram equalization contrast enhancement to display grids. For example, to display the *dogs* grid with this enhancement, issue the following command.

```
/***Display grid using stretched grayscale equal area.

Arcplot: gridpaint dogs value equalarea # gray
```

Panchromatic Display Using ArcView

Linear contrast stretch can be changed in ArcView by using the Legend Editor. Experiment with the following steps in ArcView.

1. Start ArcView and create a new view. Select Add Theme from the View menu. Double-click on *dogs.bil* to add it as a theme.

↝ **NOTE:** *Verify that Image Data Source is selected as the Data Source Type rather than Feature Data Source.*

2. Click the check box next to the theme in the legend to display the image. Double-click the image legend to open the Legend Editor.

3. Select the default for a min/max linear contrast stretch. (Additional discussion of contrast stretching in ArcView appears in the "Color Display" section later in this chapter.

Grid Display

Both the ARC/INFO GRID extension or the ArcView GIS Spatial Analyst extension allow you to work with ARC/INFO grids. In GRID, you can convert an image to a grid by using the ARC IMAGEGRID command. In ArcView, you can convert images to grids by using the Convert To Grid option under the Theme menu while the image theme is active in the view.

Grid Display Using ARCPLOT

You can display grids using the ARCPLOT commands described in the following table.

ARCPLOT command	Description	Color specified by
GRIDPAINT	Display grid as grayscale or colored grid. Color is not controlled by current shade set.	grayscale, nominal colors/remap table, color map
GRIDSHADES	Display grid cells using color shades.	shade set/remap table
GRIDQUERY	Display grid cells that satisfy a logical expression.	shade set/remap table
GRIDCOMPOSITE	Display three grids as RGB or HSV color image.	RGB and HSV color models

Grayscale Display of a Continuous Grid

First, try displaying a grid in grayscale using linear stretch contrast enhancement. The *emidalat* file used in the commands below is an elevation grid of an area near Emida in northern Idaho. Elevations range from 855m to 1337m.

```
Arcplot: display 9999 4/***Full-screen X-Windows canvas.

Arcplot: mapextent emidalat /***Scale the canvas to fit the grid
extent.

Arcplot: gridpaint emidalat value linear # gray /***Linear stretch
contrast enhancement.
```

The dark areas represent lower elevations or valleys, and the light areas represent hills. The display, although dramatic, does not offer information on specific elevations.

Color Assignment to Ranges of Grid Values

You can use a remap table to assign different colrs to elevation ranges. The *emida.rmt* remap table is a text file, and contains the following data.

```
#Remap table for elevation grid called emidalat.

#Elevation range : remapped value for that range.

855 900 : 1 #elevations from 855 to 900 will be remapped to 1

900 1000 : 2 #elevations from 900 to 1000 will be remapped to 2

1000 1100 : 3 #elevations from 1000 to 1100 will be remapped to 3

1100 1200 : 4 #elevations from 1100 to 1200 will be remapped to 4

1200 1300 : 5 #elevations from 1200 to 1300 will be remapped to 5

1300 1400 : 6 #elevations from 1300 to 1400 will be remapped to 6
```

The pound sign (#) indicates a comment; everything after the sign is ignored by ARCPLOT. The output values of 1 to 6 are used as color codes for the elevation zone display. The GRIDPAINT program has the following nominal color assignments: 0 black, 1 white, 2 red, 3 green, 4 blue, 5 cyan, 6 magenta, 7 yellow, 8 orange, 9 light green, 10 pale green, 11 pale blue, 12 dark blue, 13 pale red, 14 light gray, and 15 dark gray. Now try using the remap table with the GRIDPAINT command.

```
Arcplot: gridpaint emidalat value emida.rmt /***Paint the emidalat
grid using the remap table.
```

Color symbols can be customized in two ways. The first is to use a color map file with the GRIDPAINT program. A color map file is an ASCII file with a *.cmf* extension. The file includes an index number and corresponding RGB values for each color symbol. Try the following command line:

```
Arcplot: gridpaint emidalat value emida.rmt # emida.cmf
```

A new elevation zone map is now displayed. The new shade symbols on the map are defined by *emida.cmf*, which contains the lines listed below.

```
#Color map file for the emidalat elevation grid.

#Remapped class red intensity green intensity blue intensity

1 0 102 0

2 0 255 128

3 255 255 0

4 255 140 0

5 255 0 0

6 148 0 212
```

Every line in *emida.cmf* starts with an index number, which corresponds to the symbol number in the remap table. Following the index number are red, green, blue (RGB) values. Information on color specification appears in the "Color Display" section later in this chapter.

The second method of customizing color symbols for grid display is to use the GRIDSHADES command and specify a shade set. Try the following:

```
Arcplot: shadeset colornames.shd

Arcplot: gridshades emidalat value emida2.rmt
```

The shade set *colornames.shd* offers 129 color symbols and is very popular with ARCPLOT users. By choosing the color symbols first from *colornames.shd*, you can then specify those symbols in the remap table *emida2.rmt*, as illustrated below.

```
#Remap table for the emidalat elevation grid.

#This remap table assumes the current shade set is colornames.shd.

#

#Elevation range : remapped value for that range.

855 900 : 61 #61 from colornames.shd is dark green.

900 1000 : 68 #68 from colornames.shd is spring green.

1000 1100 : 83 #83 from colornames.shd is yellow.

1100 1200 : 105 #105 from colornames.shd is dark orange.

1200 1300 : 110 #110 from colornames.shd is red.

1300 1400 : 126 #126 from colornames.shd is blue violet.
```

Color Assignment to Selected Grid Values

To highlight and display a single elevation zone rather than all elevation zones, use the GRIDQUERY command.

```
Arcplot: clear /***Clear the canvas.

Arcplot: gridpaint emidalat value linear # gray /***Linear stretch
contrast enhancement.

Arcplot: gridnodata transparent /***Any cells in grid query that are
not selected will be clear.

Arcplot: gridquery emidalat value emida2.rmt # value > 1200 /***Color the elevations above 1200m.
```

You can also use the GRIDCOMPOSITE command to display grids using either the RGB or HSV color models. (See the "Color Display" section for further information on color models.)

Grid Display Using ArcView GIS Spatial Analyst

The Spatial Analyst extension for ArcView GIS offers capabilities for grid display similar to those of the ARC/INFO GRID module.

Grayscale Display of a Continuous Grid

You can always display a grid in ArcView by adding it as an image theme and using the linear contrast enhancement available through the Legend Editor. If you have the Spatial Analyst extension, you can also add the grid as a grid theme which allows for more flexibility in assigning colors to ranges of grid values. In the following exercise, the grid of elevation values (*emidalat*) is displayed as a grayscale grid wherein the higher elevations are lighter grays. Next, colors are assigned to elevation ranges, and all elevations above 1200m are highlighted in yellow on a grayscale display.

1. Start ArcView GIS and load the Spatial Analyst extension.

2. Choose Views in the project window, if it is not already highlighted. Click on New.

3. Select Add Theme from the View menu. Choose Grid Data Source as the Data Source Type, and double-click *emidalat* to add it as a theme. Click the check box next to *emidalat* to display the theme in the view. The elevation grid *emidalat* is a continuous (or floating point) grid, meaning that it represents a continuous surface. What you see is the display of a continuous grid using the default color ramp (red monochromatic) and the default classification method (equal interval with 9 classes). If you wish to change the symbols or the classification, access the Legend Editor.

4. Now use the Legend Editor to change the grid's color assignments. Double-click *emidalat* to open the Legend Editor.

Legend Editor for continuous grid.

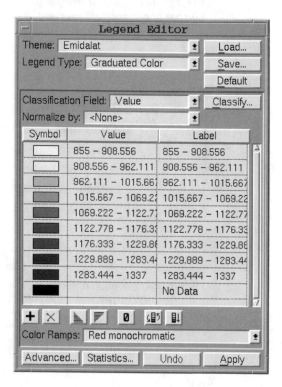

5. There are 31 choices in the Color Ramps listing. Select Gray monochromatic and click on Apply. The view is similar to the linear grayscale display using the GRIDPAINT command in ARC/INFO GRID.

Color Assignment to Ranges of Grid Values

The Legend Editor can be used to assign colors to a range of elevation values. In the next exercise, colors are assigned to six elevation zones.

1. Click on Classify in the Legend Editor to open the Classification menu.

Classification menu for assigning number of elevation classes and colors.

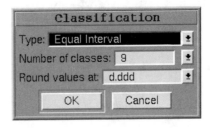

2. Select 6 as the number of classes and click on OK. There are now six classes in the Legend Editor. However, the elevation ranges in the Value column still contain odd numbers. Click the first value, type in 855-900, and press Enter. Continue to change the other values to 900-1000, 1000-1100, 1100-1200, 1200-1300, and 1300-1400.

Reclassified elevation ranges.

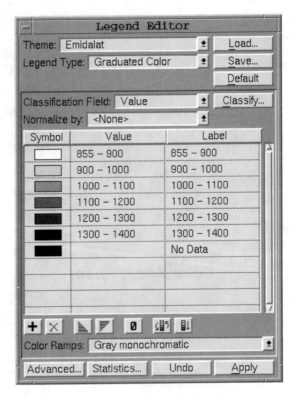

3. Apply the new grid display rule by clicking on the Apply button in the Legend Editor. The elevation grid will be displayed in the six elevation zones you chose, which is similar to the use of a remap table in ARC/INFO GRID.

4. The two color ramp choices ArcView GIS called Elevation #1 and Elevation #2 are based on the color schemes commonly used in topographic mapping. Experiment with the options.

Sometimes you may wish to change a color symbol in an otherwise satisfactory color ramp option in ArcView GIS. Using Elevation #2, for example, the color symbols look very similar at the lower and higher ends. For that reason, you may want to change the color symbol for the lowest class (855-900) to light green. Double-click on the symbol for the lowest class in the Legend Editor to open the Color Palette window. Choose a light green color from the Color Palette, and click on Apply in the Legend Editor. The color symbol has now been changed.

➡ **NOTE**: *You can also use Reclassify in the Analysis menu to perform the task of color assignment to ranges of grid values. Reclassify, however, saves the result of reclassification to a new integer grid.*

Color Assignment to Selected Grid Values

The two *.tif* files cited below are available at the *129.101.8.19* FTP site. (The user ID is *cguest*, and the password, *fr1g1d*.) There are two methods to assign color to selected grid values. The first method, described below, is simple and direct.

1. Verify that the *emidalat* grid theme is active.

2. Select Analysis | Map Query, and type the following expression in the expression box: *([Emidalat]>1200)*.

Map Query menu.

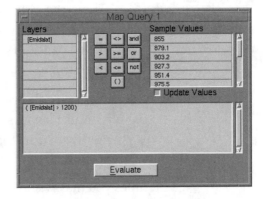

3. Click on Evaluate. This action creates a new grid named *Map Query 1*. Map Query 1 has values of 1 and 0, and no data. Cells with a value of 1 have elevations higher than 1200m, and cells with a 0 value do not.

Elevations greater than 1200m highlighted after using Map Query.

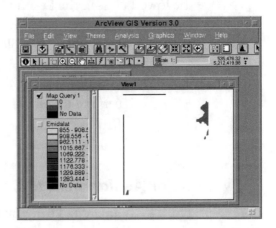

The second method begins by changing *emidalat* from a continuous grid to a discrete grid. A discrete grid has a value attribute table (*.vat*) associated with it, but a continuous grid does not. Many buttons and tools for the View document require the presence of a *.vat* file. The next exercise uses the Query Builder to highlight elevations above 1200m in red. The Query Builder uses the grid value attribute table to select grid cells based on a query. Consequently, the first step is to build a value attribute for the *emidalat* grid.

1. Verify that the *emidalat* grid theme is active.

2. Select Map Calculator from the Analysis menu.

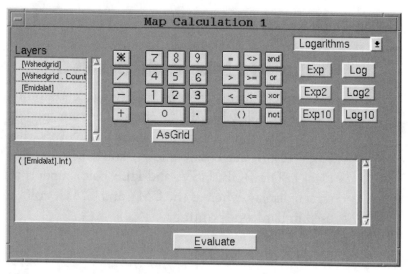

Map Calculator menu.

3. Type the following expression in the expression box: *([Emidalat].Int)*. Click on Evaluate. This action creates a new grid named *Map Calculation 1*, which is the integer version of *emidalat* with a value attribute table attached.

4. Make the *Map Calculation 1* grid active. The Query Builder icon and other icons are now visible. Click on the Query Builder button.

5. Type the following expression in the Query Builder dialog: *([Value] > 1200)*. Click on New Set. Areas with elevations higher than 1200m are highlighted in red.

To change the color assigned to selected grid cells, open the project window and select the Properties button from the Project menu. In the Project Properties menu, choose the Selection Color button to assign a new color for grid cells selected by the Query Builder. The new color is based on the HSV color model. (See the following section, "Color Display," for a discussion of the HSV color model.)

Color Display

Color Models in ARC/INFO

You can create any color by specifying a certain mixture of colors from a color model. ARC/INFO offers five color models: RGB (red, green, blue), HSV (hue, saturation, value), HLS (hue, lightness, saturation), CMY (cyan, magenta, yellow), and CMYK (cyan, magenta, yellow, black). The RGB, HSV, and HLS color models are typically used in screen display, whereas the CMY and CMYK color models are typically used in hardcopy printing.

RGB Color Model

The RGB color model is based on red, green, and blue, the three primary additive colors. This color model is standard for CRT (cathode ray tube) electronic displays. Each pixel on the computer screen consists of a triad of (phosphoric colors) red, green, and blue. Light is emitted and displayed by the video intensity at which these phosphors are projected, and the color you see on the monitor represents a mixture of red, green, and blue. White light is composed of equal proportions of the three additive primary colors at maximum intensity. Therefore, bright white pixels on the screen would be displayed with RGB values of (255, 255, 255). Black is the absence of color, and would therefore be displayed with RGB values of (0,0,0). Cyan (0, 255, 255) is created by adding green and blue light. Magenta (255, 0, 255) is created by adding red and blue light. Yellow (255, 255, 0) is created by adding red and green light.

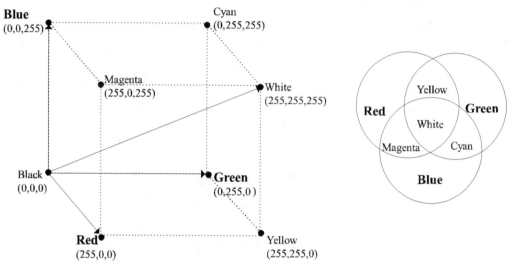

RGB color model.

You might find it difficult to visualize the colors that will be produced by various RGB combinations. For example, what does the RGB combination of (0, 127, 240) look like? Input the following statements in ARCPLOT to view the color.

```
Arcplot: display 9999 /***X-terminal canvas.

Arcplot: shadesymbol 1 /**Initialize shade symbol to 1.

Arcplot: shadecolor rgb 0 127 240 /***Specify the RGB color model
and mixture.

Arcplot: patch * /***Interactively delineate a patch box on the canvas.
```

After the first color displays, try other color combinations. With a 24-bit graphics card, you can produce over 16 million (2^{24}) different color combinations on the canvas. The LINECOLOR, MARKERCOLOR, and TEXTCOLOR commands in ARCPLOT can also be used with any RGB combination.

HSV Color Model

The HSV color model is based on the visual perception of colors. *Hue* refers to any of the colors specified in the Tektronix color standards, and is represented in the following illustration as the angle along the vertical axis. For example, 0 degrees = red, 60 degrees = yellow, 120 degrees = green, 180 degrees = cyan, 240 degrees = blue, 300 degrees = magenta, and 360 degrees = red.

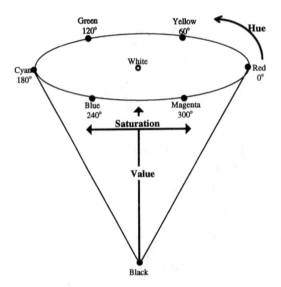

HSV color model.

The *saturation* component determines color purity, or how "washed out" a color appears. Grays have zero saturation, while pure hues have a saturation of 100. Finally, the *value* component determines color brightness. A value of zero appears black, while a value of 100 at the top of the cone is at maximum intensity. Input the following statements in ARCPLOT to experiment with the HSV color model.

```
Arcplot: clear /***Clear the canvas display.
```

```
Arcplot: shadecolor hsv 50 100 100 /***Specify the HSV color model
and mixture.
```

```
Arcplot: patch * /***Interactively delineate a patch box on the canvas.
```

HLS Color Model

In the HLS color model, *lightness* replaces *value* in the HSV model. The lightness component determines color brightness. In this model, the 0 value corresponds to black, 100 to white, and 50 to pure hues.

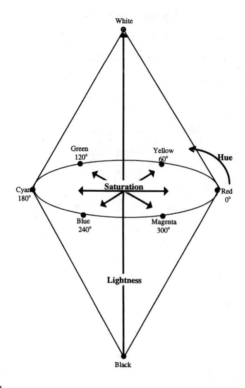

HLS color model.

Input the following statements in ARCPLOT to experiment with the HLS color model.

```
Arcplot: clear /***Clear the canvas display.
```

```
Arcplot: shadecolor hsv 100 50 50 /***Specify the HLS color model
and mixture.
```

```
Arcplot: patch * /***Interactively delineate a patch box on the canvas.
```

CMY Color Model

The CMY color model is based on cyan, magenta, and yellow, the three primary subtractive colors. This color model is the standard for color printing and photography. Printing uses pigments that absorb some of the additive primary colors. For example, cyan pigment absorbs red light and reflects the remaining light (green and blue). Thus, the viewer perceives blue-green or cyan. In a similar way, magenta pigment absorbs green light and reflects blue and red light. Yellow pigment absorbs blue light and reflects green and red light. A black pigment absorbs all visible light.

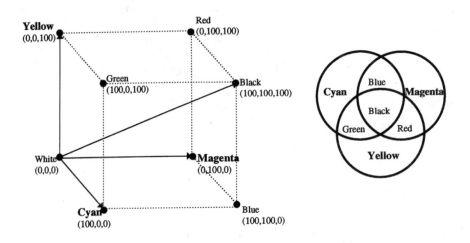

CMY color model.

The following statements in ARCPLOT use the COLORCHART command to display CMY color charts.

```
Arcplot: clear /***Clear canvas.

Arcplot: colorchart all c /***Cyan as constant color, with magenta
and yellow incremented.

Arcplot: clear /***Clear canvas.

Arcplot: colorchart all m /***Magenta as constant color, with cyan
```

and yellow incremented.

Arcplot: clear /***Clear canvas.

Arcplot: colorchart all y /***Yellow as constant color, with cyan
and magenta incremented.

CMYK Color Model

You could create the color black in the CMY model by using the combination of cyan, magenta, and yellow inks. But it would be cleaner and
simpler to add black as a fourth color. In the CMYK model, the letter
"K" represents black. (The letter "B" is not used because it could mean
blue rather than black.)

The CMYK color model is the standard for four-color process printing.
A color plate is made for each process color with different screen percentages. A printed color therefore represents a combination of different percentages of the four process colors. The screen percentage ranges
from 0 to 100%.

Consider a few basic colors and respective CMYK values. A bright red
is printed with CMYK values of (0, 100, 100, 0); a bright green, (100,
0, 100, 0); and a dark blue, (100, 100, 0, 0). To view other process color
combinations, use the SHADECOLOR command in ARCPLOT. The
statements required for viewing a CMYK value of (100, 50, 0, 50) follow.

/***Specify the CMYK color model and mixture.

Arcplot: shadecolor cmyk 100 50 0 50

Arcplot: patch * /***Interactively delineate a patch box on the canvas.

Color Translation

A color can be translated from one model to another in ARC/INFO,
but the translation is only approximate. The SHOW command in
ARCPLOT can be used for color translation. The following two state-

ments show examples of translating from CMYK to RGB, and from RGB to CMY, respectively.

```
Arcplot: show rgb cmyk 100 50 0 50 /***Output is 0, 127.5, 255.
(These are the RGB values for the color.)
```

```
Arcplot: show cmy rgb 255 255 0 /***Output is 0,0,100 (These are the
CMY values for the color.)
```

Although a nearly infinite number of colors can be created using the color models, most ARC/INFO users prefer to use the predefined sets of color symbols. The most popular shade symbol set, *COLOR-NAMES.SHD*, contains 129 named colors. Color names such as "navy blue," "sea green," or "coral" are descriptive, but if you have difficulty visualizing the colors, you can view them in ARCDOC. Every color in *COLORNAMES.SHD* is a combination based on the CMYK color model. Take a look at *$ARCHOME/symbols/colorfile* if you are interested in CMYK combinations. For example, "navy blue" has the CMYK values of (100, 100, 49.8, 0).

Color Image Display Using ARC/INFO

Color images are typically displayed using the RGB model with the IMAGE command in ARCPLOT or ARCEDIT. For example, the *color.bil* file contains the pixel values shown in the next illustration.

0	255	255	0	0	255
0	255	0	255	0	255
0	255	0	0	255	0
0	255	200	0	0	0
255	0	0	150	0	0
255	255	200	0	200	150

Pixel values inside the color.bil image.

Start by assigning the first pixel value to control red screen intensity, the second value to control green screen intensity, and the third value to control blue screen intensity.

`Arcplot: display 9999 4 /***Full-screen X-Windows canvas.`

`Arcplot: mapextent image color.bil /***Scale the canvas to fit the color.bil image.`

`Arcplot: image color.bil composite 1 2 3`

Next, try changing the red, green, and blue assignments.

`Arcplot: image color.bil composite 3 2 1 /***Third pixel value will now control red, the second green, and first, blue.`

`Arcplot: image color.bil composite 2 3 1 /***Second pixel value will now control red, third green, and first, blue.`

Now try displaying a scanned color aerial photograph of the Chena River in Fairbanks, Alaska.

`Arcplot: clear /***Clear canvas.`

`Arcplot: mapextent image chena.bil /***Scale canvas to fit the image.`

`Arcplot: image chena.bil composite 1 2 3 /***Display the image as an RGB color composite.`

Every pixel of the *chena.bil* image contains three values or bands. The statistics of these bands are read by the IMAGE command, which will then automatically apply a linear contrast stretch to the red, green, and blue screen intensities. The image statistics file has a *.stx* extension, and contains the following information for each band: band number, minimum, maximum, mean, and standard deviation. The image command uses the mean +/- 2* standard deviation to compute the upper and lower limits of the contrast stretch for each color. Therefore, if you can, adjust the red, green, or blue on the displayed image by temporarily changing the image statistics values. In the following example, the intensity of green on the display is increased.

Using a text editor, change the mean for band 2 from 125.9 to 100 in the image statistics file and then save it.

`Arcplot: &sys textedit chena.stx`

The above action will lower the contrast stretch limits so that the level of green will increase on the display. The statistics file is reproduced below.

```
1 0 234 134.3 47.1

2 0 227 125.9 52.8

3 0 218 124.6 52.1
```

To display the image with the new linear contrast stretch, use the following command line.

```
Arcplot: image chena.bil composite 1 2 3
```

Color Display Using ArcView

Compared to ARC/INFO, ArcView offers limited color model options. The HSV color model is available for features and grids and the RGB color model is available for images.

HSV Color Display of Feature Themes

1. Start ArcView GIS and create a new view by double-clicking on the New button.

2. Add the polygon theme called *compart* by selecting Add Theme from the View menu. Verify that Feature Data Source is the Data Source Type, and then double-click on *compart* to add it to the view.

3. Click on the check box next to *compart* in the Table Of Contents to display the theme.

4. To change the theme display, access the Legend Editor by double-clicking on the colored box in the legend. Edit the theme symbol with the Symbol Fill Palette by double-clicking on the colored box in the Legend Editor. Double-clicking on the paint brush icon allows you to open the Color Palette where you can select any of 60 predefined

colors for the theme. Select a color, and click on Apply in the Legend Editor to apply the new theme color.

5. You can also assign colors using the HSV model. Click the Custom button at the bottom of the Color Palette to open the Specify Color menu.

ArcView Specify Color menu.

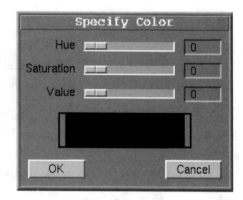

6. To change a color's Hue, Saturation, or Value in the Specify Color menu, use the slider bars or the direct input fields. Changes in color are shown at the bottom of the menu. Note that the values for Hue, Saturation, and Value range from 0 to 255. The range is not the same as the HSV color model in ARC/INFO, but the three color dimensions can be interpreted in the same way.

7. Click on OK in the Specify Color menu and Apply in the Legend Editor. The *compart* theme is displayed in the new HSV color.

RGB Color Display of Image Themes

Now try using the RGB color model with the *coldgs.bil* image.

1. Select Add Theme from the View menu. Double-click *coldgs.bil* to add it as a theme. Verify that Image rather than Feature is selected as the Data Source Type.

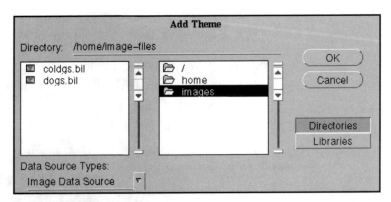

Select the image to be displayed.

2. To set the display extent, make *coldgs.bil* active, and then select View | Zoom to Themes. Click the check box next to the theme in the legend to display the image. Double-click the image legend to open the Legend Editor.

3. You can now specify the image pixel values to control red, green, and blue. Try band 1 for all three colors and then select the default button. You should see a grayscale image. Next, try band 1 to control the red, band 2 to control the green, and band 3 to control the blue, and then select the Default button. You should see a color image.

Band assignments to display a color image.

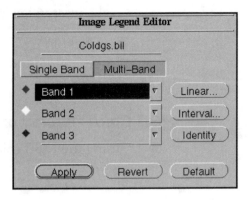

4. Pressing the Default button will cause ArcView to perform a 95 percent linear contrast stretch for each band. You can adjust the contrast stretch for each band by selecting the linear button. Try adjusting the contrast stretch for the green linear stretch such that the image appears greener than the original default color display.

    Double-click the Linear button and then shift the top of the linear stretch line in the green histogram to the left. Click on the Apply button to execute the new linear contrast stretch. Try adjusting the linear stretches for the red, green, and blue histograms until you are satisfied with the color balance.

*Adjusting the green contrast stretch
for a greener image.*

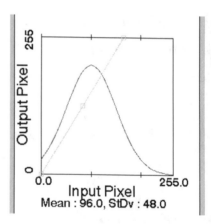

Image-Based Slide Shows

The key to fast GIS slide shows is to save the red, green, and blue screen values to a bitmapped image file. The bit-mapped image displays much faster than the original image because it is composed of fewer pixels. In ARCPLOT, you can save any display on the canvas by issuing the SCREENSAVE <slidename> command. In ArcView, you can save a view as a slide by selecting Files | Export | Windows Bitmap.

Producing a slide show can be accomplished with the use of GIS or other software. If you require a particular image format, you can use the Arc CONVERTIMAGE command to convert the bitmapped image to other image formats. In ArcView, use the Files | Export command sequence to export to a variety of image formats.

Slide Show Using ARCPLOT

There are two basic steps involved in creating ARCPLOT image slides. First, use standard ARCPLOT commands to draw to a full-screen canvas. Next, use the SCREENSAVE command to save the canvas to an image file. Repeat this process for each slide. The following command statements in ARCPLOT provide the sequence of steps involved in the two operations described above. First, create four slides from the *ak-shaded* grid, and the *towns* and *big-rivers* coverages.

```
Arcplot: display 9999 4 /***Full-screen X-Windows canvas.

Arcplot: mapextent image ak-shaded /***Scale canvas to fit image.

Arcplot: mapposition cen,cen /***Position to center of canvas.

/***Slide 1 of Alaska and then save it as bitmapped screen save.

Arcplot: image ak-shaded /***Display grayscale image to canvas.

Arcplot: linecolor cyan; arcs big-rivers /***Draw big-rivers in cyan.

Arcplot: textfont times; textquality proportional; textsize .75

Arcplot: move 9,7; text 'Alaska'

Arcplot: screensave slide1

/***Make slide 2 of Juneau.

/***Mapextent around Juneau area.

Arcplot: mapextent 1012447,834735,1316256,1243564

Arcplot: clear /***Clear canvas.

/*** Ocean background.
```

Arcplot: shadesymbol 1; shadecolor rgb 0,150,200; patch 0,0 20,20

Arcplot: gridnodata transparent /***Ocean pixels will be transparent.

/***Display grid using linear contrast.

Arcplot: gridpaint ak-shaded value linear # gray

Arcplot: markerset plotter

/*** Magenta star.

Arcplot: markersize .25; markersymbol 2; markercolor magenta

/*** Plot Juneau marker.

Arcplot: reselect towns point name cn 'Juneau';points towns

Arcplot: textsize .5; textcolor magenta; pointtext towns name

Arcplot: aselect towns point

Arcplot: screensave slide2

/***Make slide 3 of Anchorage.

Arcplot: mapextent image ak-shaded

/*** Anchorage area.

Arcplot: mapextent 52261,1063529,434835,1386092

Arcplot: clear /***Clear canvas.

Arcplot: patch 0,0 20,20 /***Ocean background.

/***Display grid using linear contrast.

Arcplot: gridpaint ak-shaded value linear # gray

Arcplot: linecolor cyan; arcs big-rivers /***Draw big-rivers in cyan.

/***Cross for marker symbol.

Arcplot: markersymbol 2;markercolor magenta; markersize .5

/*** Anchorage marker.

```
Arcplot: reselect towns point name cn 'Anchorage';points towns

Arcplot: pointtext towns name /***Anchorage text.

Arcplot: aselect towns point

Arcplot: screensave slide3

/***Make slide 4 of Fairbanks.

Arcplot: mapextent image ak-shaded

Arcplot: mapextent 107005,1496836,550610,1935192

Arcplot: clear /***Clear canvas.

Arcplot: patch 0,0 20,20 /***Ocean background.

Arcplot: image ak-shaded /***Display grayscale image.

Arcplot: linecolor cyan; arcs big-rivers /***Draw big-rivers in cyan.

Arcplot: reselect towns point name cn 'Fairbanks';points towns

Arcplot: pointtext towns name /***Fairbanks text.

Arcplot: screensave slide4
```

Next, display the slides with the simple macro named *slideshow.aml*.

```
Arcplot: &run slideshow.aml
```

Hot Link in ArcView

ArcView's Hot Link feature allows you to link a theme point, line, or polygon feature with an action by simply clicking on the feature in the view. The hot link can open a text file, an image, an ArcView document or project, or even run an Avenue script. Hot Link is a tool you can use to set up an interesting, and even dramatic demonstration project. A user can click on a map feature with a hot link and get additional information about the feature. A classic example is a real estate application wherein the GIS user selects all candidate houses in the

$75,000-$100,000 price range within one mile of a grade school and within 20 miles of a lake. Using hot links, the user could then click on any house that was selected based on this query to see a floor plan and photo of the house.

In the following exercise, hot links from the *compart* theme are used to link to digital orthophotos (*compart44.tif* and *compart55.tif*). Compartments are management units delineated for forestry applications such as silvicultural planning.

The first step in hot linking is to prepare the "link to" objects. In this example, orthophotos to be linked with two compartments have been clipped for the area around each compartment polygon. Information on clipping grids using polygons appears in Chapter 10, "Grid Management, Clipping, and Resampling."

The second step is to add the "link to" object name in the theme attribute table. You can then direct ArcView to use a particular field in the attribute table for hot linking.

1. Create a new view by selecting Views in the project window and clicking on the New button.

2. Select View | Add Theme and then double-click on *compart* to add it as the theme.

3. Add the "link to" object name in the theme attribute table. From the Theme menu, choose Table to open the polygon attribute table. Note that the field called *Links-to* is the column at the extreme right of the table.

4. The link to object name for compartment 44 (*compart44.tif*) has already been entered for you. You must now enter the link to object name for compartment 55. Select Start Editing from the Table menu and choose the Change Cell Values button. Select the cell to the right of

55 in the *compart* field. Type in *compart55.tif* as the link to object name for that compartment polygon. Press <Enter> to complete the entry.

Shape	Area	Perimeter	Compart#	Compart-id	Compart	Links-to
Polygon	44822732.000	42088.367	2	1	44	compart44.tif
Polygon	1.844976e+08	81628.219	3	2	53	
Polygon	5220891.000	9277.292	4	3	53	
Polygon	1.434286e+08	54048.816	5	4	55	compart55.tif
Polygon	519255.906	3423.093	6	5	53	
Polygon	498091.875	3400.496	7	6	53	
Polygon	1676213.625	5179.922	8	7	53	
Polygon	228407.094	3449.856	9	8	53	
Polygon	2537529.500	10586.536	10	9	53	

Entering compart55.tif as a "link to" object for compartment 55.

⟶ **NOTE:** *If* compart44.tif *and* compart55.tif *are not in the same workspace as the* compart *coverage, you need to provide the path to the TIFF files in the input (e.g., /home/ image/compart44.tif).*

5. At this point, you have defined the "links to" in the polygon attribute table. Select Table | Stop Editing and save the edits.

6. Next, define the hot links as a theme property. Open the view again so that the compartment polygons are visible. Select Theme | Properties. Scroll down below the Display icon and select the Hot Link icon to display the hot link property options.

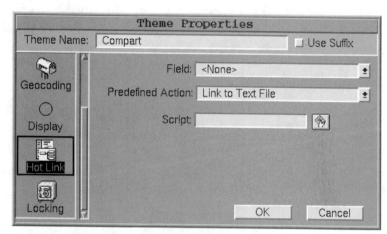

Select the Hot Link icon to define the hot link property.

7. Select Links-to for the Field, and Link to Image File for the Predefined Action. Click on OK. The setup for hot links is complete.

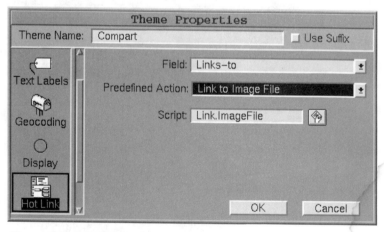

Theme Properties dialog box.

8. Determine where the polygons with links to images are located in the view. Verify that the *compart* theme is checked and active. Enter the query *compart = 44 or compart = 55* with the Query Builder.

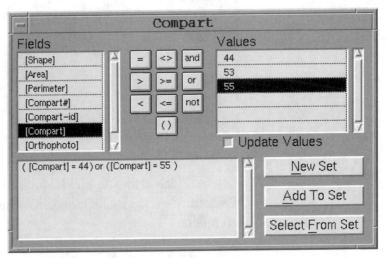

Query Builder for selecting compartments 44 and 55.

9. Click on New Set to select Compartments 44 and 55 as the net set of selected polygons. The selected polygons will be highlighted in the view.

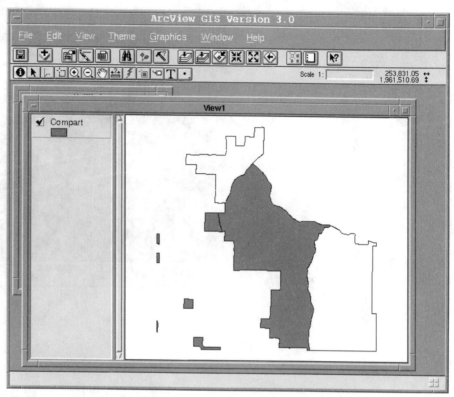

Compartments 44 and 55 highlighted after using the Query Builder.

10. Test the hot links. Click the Hot Link tool (the lightning bolt icon), place the cursor inside a highlighted polygon, and click. These actions should open the orthophoto linked to that compartment.

By default, the image will be scaled to fit inside the hot link window. Consequently, its shape may be distorted. Check the scale image box to retain the true shape of the image. Then drag a corner of the hot linked image to enlarge the orthophoto and display more detail.

Image of the hot linked polygon.

2

Surfaces

Three-dimensional (3D) surfaces and shaded relief images can be displayed in a GIS. If you have a TIN license, you can develop these 3D perspective views and shaded relief images using ARC/INFO. The Spatial Analyst extension for ArcView GIS can also be used to develop shaded relief images. As of the time of this writing, ESRI had announced the upcoming release of an ArcView GIS extension for developing 3D perspective views.

Files from the companion CD to be used in exercises and as references are listed in the table below.

emidalat	Grid of elevation values near Emida, northern Idaho.
emidashade	Shaded relief grid produced using the ARC HILLSHADE command.
emida.lut	INFO lookup table for assigning colors to specific elevation zones.
emidacont	Line coverage of contours for the Emida area.
ak-shaded	Shaded relief grid of most of Alaska produced using the ARC HILLSHADE command.
big-rivers	Line coverage of major rivers of Alaska.
towns	Point coverage of cities, towns, and villages in Alaska.
iditarod	Line coverage of the Iditarod mushing trail in Alaska.
surfaces.doc	ARC/INFO documentation file.

Perspective Views

Three parameters are typically adjusted when viewing surfaces: viewing azimuth, viewing angle, and surface z-scale. The azimuth is the direction from the surface to the observer and ranges from 0 to 360 degrees in a clockwise direction. The following figures show perspective views of most of Alaska from three different viewing azimuths.

Perspective with viewing azimuth set to 215 (viewed from southwest).

Perspective with viewing azimuth set to 180 (viewed from south).

Perspective with viewing azimuth set to 315 (viewed from northwest).

The viewing angle, always between 0 and 90 degrees, is measured from the horizon to the altitude of the observer. A view angle of 90, for example, displays the surface from directly above, while a view angle of 0 would show you the surface in profile. The following figures are perspective views of most of Alaska from three different viewing angles.

Perspective with viewing angle set to 10 degrees.

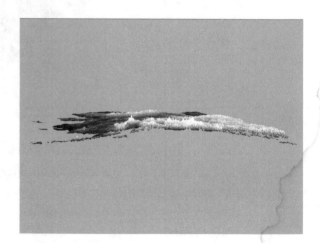

*Perspective with
viewing angle set
to 30 degrees.*

*Perspective with
viewing angle set
to 50 degrees.*

You can also change the vertical exaggeration of a viewing surface by changing the surface z-scale value. The larger you set the surface z-scale, the more rugged the surface will appear. For example, the following figures are perspective views of most of Alaska with increasing surface z-scale values.

Perspective view of most of Alaska with surface z-scale set to 10 factor.

Perspective view of Alaska with surface z-scale set to 25 factor.

Perspective view of Alaska with surface z-scale set to 50 factor.

⟶ **NOTE:** *You need an ARC/INFO TIN license to develop perspective views and shaded relief images from elevation grids. Look for a TIN license in the $ARCHOME/sysgen/ license.dat file or check with your system administrator.*

In the next section, selected perspective views of an area near Emida, Idaho, are developed which correspond to the following topographic map.

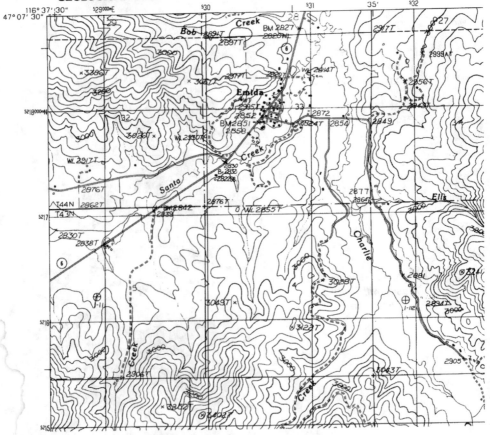

Emida quadrangle, northern Idaho.

In the following exercise, you will work with a lattice (grid) file called *emidalat*. The ARC/INFO command DEMLATTICE was used to convert the 7.5-minute DEM to the lattice file, which is essentially a regular grid filled with elevation values at 30m intervals. The second data file is a shaded relief grid called *emidashade*. The grid was created by using the ARC/INFO HILLSHADE command and *emidalat* as input. HILLSHADE uses the azimuth and altitude parameters, which are discussed in the section, "Changing the Viewing Azimuth and Altitude."

Perspective Views Using ARCTOOLS

Simple 3D View

1. Invoke ARCTOOLS mapping tools with the following command.

```
Arc: arctools map
```

2. Create a new view. In the Map Tools menu, select View | New to open both the Theme Manager and Add New Theme menus.

The Add New Theme menu is used to add and display a new theme. The Theme Manager maintains the property sheet for each theme, which identifies the theme's data source and symbology. The Theme Manager also controls the themes to be drawn and the order in which they are drawn.

3. The next step is to load a digital elevation grid as a surface to be viewed. In the Map Tools menu, select View | Load Surface.

↪ **NOTE:** *A surface must be loaded for 3D views. The surface remains active until a new one is loaded.*

4. In the Surface Properties menu, click on the Lattice button as the Surface type and input *emidalat* as the Surface. Click OK.

Surface Properties menu.

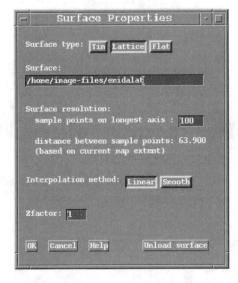

5. At this point, information about the surface lattice must be supplied. In the Add New Theme menu, select Grid for categories and Grid for theme class.

Add New Theme menu.

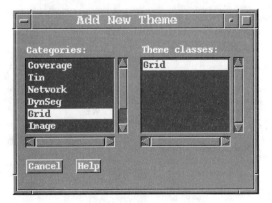

6. In the Grid Property Sheet dialog, enter *emida_elev* in the Identifier field and *emidalat* for the Data source. Click the Continuous button, and then click on OK.

Grid Property Sheet dialog box.

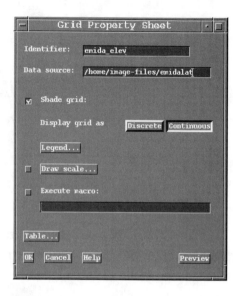

~ **NOTE:** *The Identifier is a name assigned by the user for easy identification and for use in the Theme Manager menu. The Identifier may be different from the name of the coverage or file.*

7. In the Theme Manager menu, verify that *emida_elev* is highlighted under Themes. Click the right arrow button to transfer the theme to the Draw list, and then click on the 3D button.

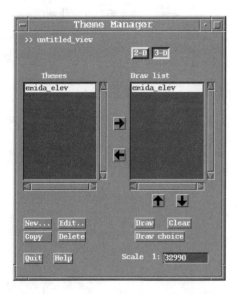

Theme Manager dialog box.

8. Click the Draw button in the Theme Manager dialog. A 3D display will appear in the ARCTOOLS window.

3D display.

Changing the Viewing Azimuth and Altitude

To change the azimuth and altitude of the 3D view from above, take the following steps.

1. In the Map Tools menu, select View | Orient Camera to open the Camera Orientation dialog box.

Camera Orientation dialog box.

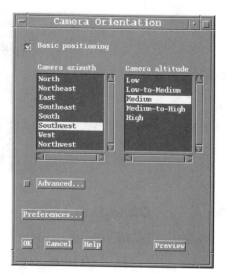

2. Basic positioning in the Camera Orientation dialog box offers the listed choices for camera azimuth and camera altitude. Select azimuth and altitude values from the lists, and click on OK.

3. Click the Clear button beneath the Draw list in the Theme Manager menu to clear the previous surface viewing parameters. Click the Draw button to drawthe surface using the new parameters. A new 3D view is displayed in the ARCTOOLS canvas.

New 3D display.

Changing Other Viewing Parameters

In addition to azimuth and altitude, other viewing parameters can be adjusted and changed. These parameters are accessed via the Advanced Camera Orientation menu and the Grid Property Sheet dialog box. Both options are discussed below.

Advanced Camera Orientation

The Advanced Camera Orientation dialog can be accessed via the Advanced button in the Camera Orientation menu. This dialog box offers parameters relating to the camera position, target position, and viewfield angles. You can use the icons or direct input menu to specify parameters.

*Advanced Camera Orientation
dialog box.*

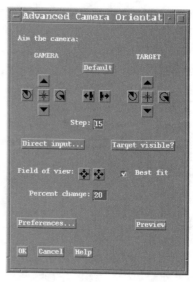

The icons are easy to use. Pressing the clockwise button for the camera position, for example, will move the azimuth clockwise around the target. Pressing the up arrow button for the target position will raise the target height. The Default button resets the camera position to the southwest and the target position to the center. If you click the Best fit buttons to reduce or increase the viewfield angle, ARCTOOLS will choose the largest surface area that can be viewed.

The Direct Camera/Target Input dialog allows you to input explicit input values. You can enter exact coordinates for the camera and target position, as well as values for the azimuth, tilt, and distance of the camera to the target. You can also specify the field of view by entering the above, below, left, and right angles.

Direct Camera/Target Input menu.

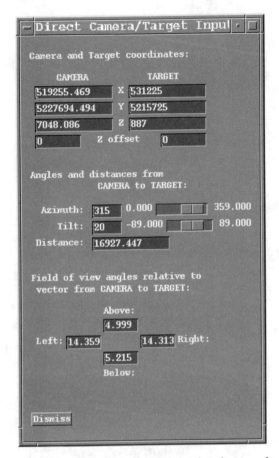

The Preferences button in both the Camera Orientation and Advanced Camera Orientation dialogs invokes the Camera Preferences dialog. This dialog allows you to experiment with parameters such as the vertical exaggeration factor and the projection method for viewing. The default exaggeration factor is 2; a value higher than 2 results in more vertical exaggeration in 3D viewing. The default projection method is perspective, which simulates a standard camera and is similar to what a human eye might see. Alternatively, the panoramic method simulates a panoramic (wide-angle) camera.

The Preview button in the Advanced Camera Orientation menu has two options. You can choose a surface mesh and its respective color symbol, or the orientation graphs depicting the camera position relative to the target in previews. A preview therefore lets you see the 3D setup using a surface mesh and/or orientation graphs.

Grid Property Parameters

- *Grid Property Sheet.* The sheet is accessed by clicking on the Edit button in the Theme Manager menu. A grid can be displayed as a discrete or continuous surface. A discrete surface is comprised of clusters of pixels with different symbols. A continuous surface has a smooth appearance and is preferred by most people.

- *Discrete Legend Editor or Continuous Legend Editor.* Either editor is invoked by clicking on the Legend button, depending on the surface type for display. Both menus allow you to select symbology and classification. The symbology options include Rainbow, Grayshade, Colorramp, and Browse. Rainbow is the default option. The classification options cover the number of classes and the classification method. With each change in symbology or classification, you can click the preview button to preview the change in the data display. (Detailed coverage of the Legend Editor appears in the next section.)

- *Draw Scale button.* Clicking on this button permits you to specify the minimum and maximum map scale to be displayed. The default is to let ARCTOOLS choose an optimal scale based on the ARCTOOLS window size.

- *Table button.* Clicking on this button provides access to options on the grid's attribute items.

Hypsometric Coloring and 3D Views

Hypsometric coloring (or layer tinting) is a common method of mapping land surfaces, especially at a small scale. The land surface is divided into a series of elevation zones, and each zone is assigned a color. A traditional color scheme for hypsometric coloring includes blue for water, and green, yellow, and brown for land areas ranging from low to high elevations.

The Legend Editor provides options for applying hypsometric coloring to 3D views. The first step in this process is to prepare a lookup table in ARC before you start ARCTOOLS. The lookup table must contain value and symbol fields; a meaning field is optional. The value field contains the class breaks in elevation, and the field symbol contains the symbol numbers. The meaning field can be used to store class descriptions.

The *emidalat.lut* file is a lookup table. Elevations in the *emidalat* grid file range from 855 to 1337m. Therefore, in *emidalat.lut*, elevations are grouped into five classes: less than 900, 900-1000, 1000-1100, 1100-1200, and greater than 1200. The color symbols, chosen from the *COLORNAMES.SHD* shade set, include 61 (dark green), 73 (green yellow), 83 (yellow), 105 (dark orange), and 110 (red). Experiment with hypsometric coloring using ARCTOOLS by following the steps below.

1. Start the ARCTOOLS map tools with the following command.

    ```
    Arc: arctools map &
    ```

2. Create a new view for the hypsometric coloring. In the Map Tools menu, select View | New to open the Theme Manager. Next, select Add New Theme. Before you use these dialogs, you must load a 3D surface. Therefore, select View in the Map Tools menu again, and then select Load Surface.

3. At this juncture, you should specify information about the viewing surface. In the Surface Properties menu, click on the Lattice button as the surface type and key in *emidalat* as the surface. Click OK. In the Add New Theme menu,

select Grid for Categories and Grid for Theme Class. Next, in the Grid Property Sheet dialog, key in *emida_elev* for the Identifier and *emidalat* for the Data Source. Click the Discrete button, followed by the Legend button.

4. Now you can assign shade colors to elevation zones. Specify that you want to shade discrete classes by clicking the Symbolset button in the Discrete Legend Editor dialog.

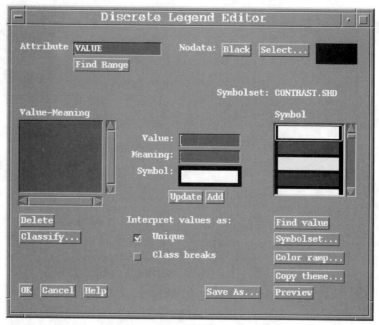

Discrete Legend Editor dialog box.

5. In the Select a Symbolset File dialog, select the shade set called COLORNAMES.SHD from the list. If you wish, use ARCDOC to view the colors available in the shade set. Click on OK.

6. Now you can specify which lookup table to use in shading the elevation zones. Click the Classify button in the Discrete Legend Editor. Select Lookup_Table in the Classification menu, and then click on Apply.

Classification menu.

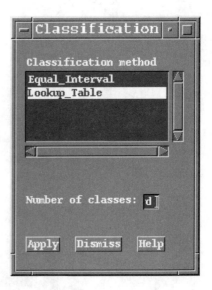

7. Select the *emida.lut* lookup table from the Select A File menu. Click on OK.

8. In the Discrete Legend Editor, click on each class under Value-Meaning and view the color symbol chosen for the class. If you wish to change a color symbol, select a color from the Symbol list and click on the Update button. If you wish to change a class break, enter a new break in the Value field and click Update. Every time you make a change in symbology or classification, click the Preview button so that you can see the change in display. When you are satisfied with the classification and symbology, click the Save As button to save the lookup table for future use.

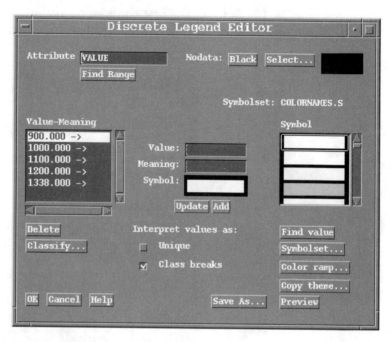

Discrete Legend Editor dialog box.

9. Once you are satisfied with the classification and symbol-ogy, you can use it in the surface display. Click OK in the Grid Property Sheet dialog. In the Theme Manager menu, verify that *emida_elev* is highlighted under Theme. Click the right arrow button to transfer the theme to the Draw list. Click the 3D button, and then select the Draw button in the Theme Manager menu. A 3D display based on the selection of elevation zones and color symbols will appear on the ARCTOOLS canvas.

Draping Images and Coverages

You can achieve dramatic displays by draping thematic data such as land cover, vegetation, roads, or other surface features onto 3D views. The following steps show you how to achieve simple 3D draping in ARCTOOLS.

1. Refer to the first three steps in the "Advanced Camera Orientation" section to display a 3D view.

2. A line coverage called *emidacont* contains contour lines for the 3D display area. In this exercise, you will drape *emidacont* onto the 3D view. In the Theme Manager menu, select the New button to open the Add New Theme menu.

3. Select Coverage from the Categories list and Line from the Theme Classes list.

4. In the Line Theme Properties menu, key in *emida_cont* as the Identifier and *emidacont* as the Data Source. Select a symbol from the Symbolset list and then click OK.

Line Theme Properties menu.

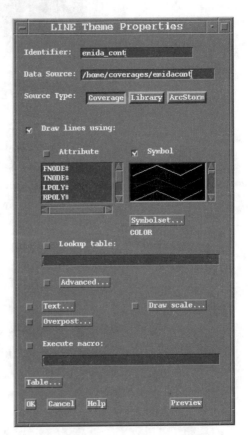

5. In the Theme Manager menu, select *emida_cont*. Click the right arrow button and then the Draw Choice button. The 3D view is now draped with contour lines in the ARC-TOOLS window.

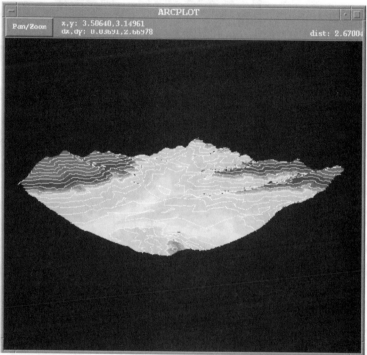

3D draping.

Options for 3D Draping

- The Advanced button in the Line Theme Properties menu invokes the Discrete Legend Editor, which is used to select classification and symbology for draping coverages.

- The Text button in the Line Theme Properties menu opens the Coverage Text Properties menu, which is used to specify font family, style, size, and color for text draping. The Overpost button is used to detect and resolve text conflicts.

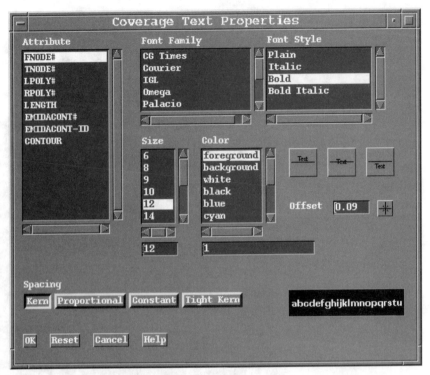

Coverage Text Properties menu.

- The Draw Scale and Table buttons in the Line Theme Properties menu are similar to those found in the Grid Property Sheet discussed earlier. The Draw Scale button allows you to specify the minimum and maximum mapping scales to be displayed. The Table button provides access to the theme's attribute table.

Perspective Views Using ARCPLOT Commands

You can also generate perspective views from the ARCPLOT prompt. This method has several advantages over ARCTOOLS. First, because perspective views of large grids require substantial computer processing time, you can submit large perspective viewing jobs to be performed as an off-hour batch job. Submitting batch jobs is discussed in Chapter 13, "System Integration." Second, you can use the Arc Macro Language to customize perspective viewing. In addition, some users prefer ARC-PLOT's command line approach over the menu-based ARCTOOLS approach.

The next section introduces the basics of surface viewing using ARC-PLOT commands, but only scratches the surface in terms of the options available for surface viewing at the ARCPLOT prompt. The following exercises will show you how to generate the three types of perspective views covered under the ARCTOOLS discussion: simple 3D surface, hypsometric colored surface, and draping on a surface.

Simple 3D Surface

A fishnet mesh is often used to preview surfaces because it displays quickly. Once you have determined the best view angle, view direction, surface z-scale, and other viewing parameters with a draft fishnet display, you can use slower surface display techniques of greater display quality. A fishnet mesh perspective is developed in ARCPLOT below.

```
Arcplot: display 9999 4 /***X-Windows full-screen canvas.

Arcplot: mapextent emidalat /***Scale canvas to fit the elevation
grid.

/***Generate a surface from the emidalat grid.

/***Generate the surface from the elevation grid.

Arcplot: surface lattice emidalat

Arcplot: surfacedefaults /***Default viewing angle 30, azimuth 215.

Arcplot: surfaceinfo /***Check out surface viewing parameters.

/***Use fishnet for relatively quick draft viewing.

Arcplot: clear /***Clear the canvas.

Arcplot: surfacedrape mesh /***Drape the fishnet on the surface.
```

By default, surfaces are viewed from the southwest (aspect of 215) at a 30 degree angle. You can easily change these values using the ARC-PLOT SURFACEOBSERVER command. An example follows.

```
/***Change observer to view from south and to 50 degree viewing angle.

Arcplot: surfaceobserver relative 180 50 #

/***Drape the mesh on the surface using new perspective viewing
rules.

Arcplot: surfacedrape mesh
```

You can increase the exaggeration of the relief by increasing the surface z-scale value, as shown below.

```
/***Increase vertical exaggeration.

/***Elevation will be exaggerated by a factor of 5.

Arcplot: surfacezscale 5 constant

Arcplot: clear

/***Drape the mesh on the surface using new perspective viewing
rules.

Arcplot: surfacedrape mesh
```

Draping Features on a Surface

With the SURFACEDRAPE command, you can use any ARCPLOT drawing commands such as GRIDPAINT, GRIDSHADES, IMAGE, ARCS, POINTS, POLYGONSHADES, and so on to drape features on top of a surface. In the following example, the elevation grid is draped as a linear contrast, stretched grayscale image. The *emidacont* line coverage is then draped on the surface.

```
Arcplot: clear /***Clear the canvas.

/***Drape the grid as a linear stretched grayscale image.

Arcplot: surfacedrape gridpaint emidalat value linear # gray

/***Drape contours on surface as brown arcs.
```

```
/***Set the current linecolor to brown.
```

```
Arcplot: linecolor brown
```

```
/***Drape the arcs from emidacont.
```

```
Arcplot: surfacedrape arcs emidacont
```

Hypsometric Coloring

The *emida.lut* INFO table is designed to be used with the shadeset COL-ORNAMES.SHD to assign specific colors to each elevation zone. You can display the lookup table with the following commands.

```
Arcplot: clear /***Clear the canvas.
```

```
/***List the lookup table that assigns colors to elevation ranges.
```

```
Arcplot: list emidalat.lut
```

Record	VALUE	SYMBOL
1	900.000	61
2	1000.000	73
3	1100.000	83
4	1200.000	105
5	1400.000	110

Tables can be used to change color assignments. For example, the current version of *emidalat.lut* assigns all cells with an elevation of less than 900 to the color symbol of 61. Try using the lookup table to produce a hypsometric colored view of the surface.

```
/***Specify the shade set the lookup table was designed to use.
```

```
shadeset colornames
```

```
/***Drape elevations over the surface using the lookup table.
```

```
surfacedrape gridshades emidalat # emidalat.lut nowrap
```

Shaded Relief

Like 3D viewing, shaded relief helps users to visualize three-dimensional surfaces. Shaded relief is also known as hill shading or, simply, shading. Shaded relief attempts to simulate what people see in the real world; that is, interaction between sunlight and land surface features. For example, a hill slope directly facing incoming light will be very bright, but a slope opposite to the light will be very dark. Real world experience helps viewers recognize the shape of land surface features on a shaded relief display.

Azimuth and altitude influence the visual effect of shaded relief. In a shaded relief display, azimuth is the direction of the incoming light, and altitude is the angle of the incoming light as measured above the horizon. ARC/INFO's default azimuth and altitude values are 315 degrees (northwest) and 45 degrees, respectively.

Shaded Relief Display Using ARCTOOLS

Similar to 3D viewing, shaded relief can be easily displayed using ARC-TOOLS. Take the following steps to experiment with shaded relief display.

1. Start ARCTOOLS map tools as follows:

   ```
   Arc: arctools map &
   ```

2. In the Map Tools menu, select View | New. The Theme Manager and Add New Theme menus will open. Before you use these menus, however, you must load a surface.

3. In the Map Tools menu, select View | Load Surface. In the Surface Properties dialog, click on the Lattice button as the Surface type and key in *emidashade*, including the directory path, as the Surface. Click on OK.

Surface Properties menu.

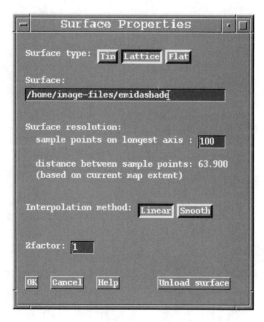

4. Specify that the surface contains continuous elevation values. In the Add New Theme dialog, select Grid for Categories and Grid for Theme Class. In the Grid Property Sheet dialog box, key in *emida_shade* for the Identifier and *emidashade* for the Data Source. Click on the Continuous button.

5. Assign a continuous grayscale shading to the grid. Click on the Legend button. In the Continuous Legend Editor, select Grayshade from the Symbology list and click on OK.

Continuous Legend
Editor.

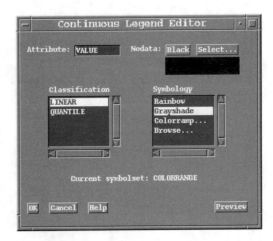

6. To display the grayscale shaded relief grid, click on OK in the Grid Property Sheet dialog box. In the Theme Manager menu, verify that *emida_shade* is highlighted under Theme. Click the Right Arrow button to transfer the theme to the Draw list. Click on the 3D button, and then the Draw button in the Theme Manager menu. A 3D display will appear on the ARCPLOT canvas.

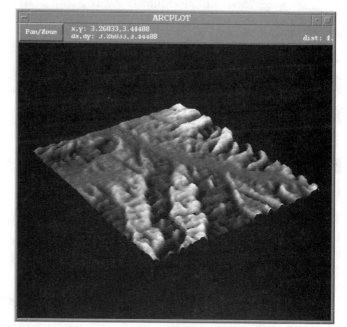

Shaded relief map.

Shaded Relief Display Using ARCPLOT

If you prefer using ARCPLOT commands, you can easily display a shaded relief grid using either linear or equal area contrast enhancements. The following documentation file (*surfaces.doc*) displays a shaded relief grid of most of Alaska. The major rivers are draped in blue, and the Iditarod mushing trail in red.

```
Arcplot: display 9999 4 /***Full-screen canvas.

/***Scale canvas to fit the shaded relief grid.

Arcplot: mapextent ak-shaded

/***Grayscale linear contrast enhancement.

Arcplot: gridpaint ak-shaded value linear # gray

/***Grayscale equal area contrast enhancement.
```

```
Arcplot: gridpaint ak-shaded value equalarea # gray

/***Drape the major rivers as blue arcs.

Arcplot: linecolor blue; arcs big-rivers

/***Drape the Iditarod trail as red arcs.

Arcplot: linecolor red; arcs iditarod
```

Shaded Relief Display in ArcView

Hillshade is one of the surface analytical functions built into the Spatial Analyst extension for ArcView GIS 3.0. With the use of the extension, you can display a shaded relief in ArcView by using an elevation grid such as *emidalat* as the input data. Try experimenting with shaded relief display in ArcView. The following exercise assumes that ArcView GIS and the Spatial Analyst extension are loaded.

> ➙ **NOTE:** *If you are running ArcView on a PC, you can use Import71 to import the interchange file for* emidalat *into your PC workspace.*

Spatial Analyst extension.

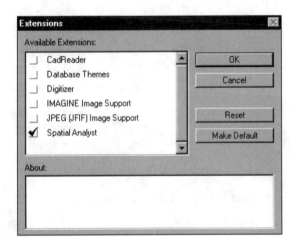

1. Select Views in the project window if it is not already high-lighted, and click on New. In the View menu, select Add Theme.

2. Choose Grid Data Source as the Data Source Type. Double-click on *emidalat* to select the elevation grid as the theme (i.e., input data).

3. Verify that *emidalat* is active (raised in the View Table of Contents), and then select Compute Hillshade from the Analysis menu.

Analysis menu.

4. In the Compute Hillshade dialog, input the azimuth and altitude values if you wish to change them from the default values of 315 and 45 degrees, respectively. Click on OK. A Hillshade theme is added to the legend.

5. Display the Hillshade theme by clicking the check box next to it in the View table of contents.

Shaded Relief theme.

✓ ***TIP:*** *To change the name of the shaded relief theme to a more meaningful one, select Properties from the Theme menu and enter a new theme name.*

Variations in Shaded Relief Display Using ArcView

The symbology for Hillshade can be changed using the Legend Editor in ArcView. Take the following steps to experiment with changing Hillshade symbology.

1. Double-click on the Hillshade theme in the legend to open the Legend Editor.

2. Click on the Advanced button to open the Advanced Options dialog box. Select the Hillshade theme as the Brightness Theme, and then set the Minimum Cell Brightness to 20 and the Maximum Cell Brightness to 80. Click on OK.

Advanced Options dialog box.

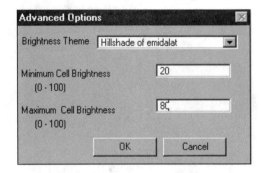

3. Click on Apply in the Legend Editor. This action changes the symbology from a range of white to black to a range of light to dark gray, resulting in a more subdued shaded relief display.

4. The Legend Editor includes a scroll list of color ramps. Gray monochromatic is the default, and will probably be suitable in most cases. Other color ramp options can also be applied to shaded relief displays.

Part II
Image Data

Scanning

In some ways, scanning is analogous to photocopying. Like photocopying, you can scan in grayscale or color and specify an output page size. A photocopier scans an image, and then prints the image to paper. A scanner scans aerial photographs, and mylar or paper maps, and then saves the image to a file.

At the time of this writing, desktop scanners generally cost less than US$1,000, while photogrammetric quality scanners ranged from US$50,000 to above US$100,000. The latter are typically used to scan metric quality aerial photographs for precise softcopy photogrammetry applications. Most of the discussion in this chapter pertains to the use of desktop scanners in non-photogrammetric GIS applications.

The files from the companion CD to be used in this chapter are listed below.

hoytmtn.tif	Scanned 7.5-minute quad of Hoyt Mountain, Idaho.
hoytmtn.tfw	Word file for scan of Hoyt Mountain, Idaho.
hoytmtn-tics	ARC/INFO coverage containing UTM coordinates for quad of Hoyt Mountain, Idaho.
scanning.doc	ARC/INFO documentation file.

Bits and Bytes

Because scanning saves an image to a file, it is important to understand how computers store binary integers. The foundation of integer storage is the *bit*, a "computer memory cell." A bit can have only two values, on or off. With a single bit, the numbers 0 and 1 can be represented as illustrated below.

$(2^0 = 1)$	
OFF	= 0
ON	= 1

With four bits, any integer ranging from 0 to 15 can be represented as shown below.

$(2^0 = 1)$	$(2^1 = 2)$	$(2^2 = 4)$	$(2^3 = 8)$	
OFF	OFF	OFF	OFF	= 0
ON	ON	ON	ON	= 15

A *byte* consists of eight bits. The maximum integer value that can be stored using a single byte is 255.

$(2^0 = 1)$	$(2^1 = 2)$	$(2^2 = 4)$	$(2^3 = 8)$	$(2^4 = 16)$	$(2^5 = 32)$	$(2^6 = 64)$	$(2^7 = 128)$	
OFF	OFF	OFF	OFF	OFF	OFF	OFF	OFF	=0
ON	ON	ON	ON	ON	ON	ON	ON	=255

Color versus Grayscale Scanning

Many scanners generate output in grayscale or color mode. In grayscale scanning, the scanner samples across a scan line for levels of gray within each sample cell. For each sample cell scanned, a value ranging from zero (for black) to 255 (for white) is stored in the output file. In color scanning mode, each cell in the scan line is sampled for levels of red, green, and blue reflected light. Thus, for each sample cell a value ranging from zero to 255 is stored for each of the three primary colors.

Because each cell contains a value ranging from 0 through 255, a grayscale scan stores one byte (8 bits) per cell. In color scanning, however, each cell must contain a value from 0 through 255 for the red, green, and blue values. Therefore, in a color scan every color requires 1 byte, and every cell requires 3 bytes.

Binary Threshold Scanning

Scanning a black and white manuscript does not require color or levels of gray. For example, scanning mylar separates, black and white CAD drawings, or black and white polygon manuscripts can be stored more efficiently by binary scanning than grayscale scanning.

In binary scanning, a threshold scan value must be specified. All cells below the threshold value are stored as zeros, and all cells above the value are stored as ones. Determining the appropriate threshold value is typically an iterative trial and error process, as seen in the following examples of binary scans from a wetlands mylar map.

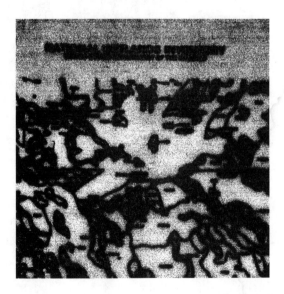

Binary scan with threshold value of 200, resulting in a very dark image.

Binary scan with threshold value of 128, resulting in a lighter image.

Binary scan with threshold value of 50, resulting in a clear scan.

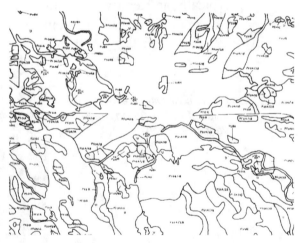

The advantage of binary threshold scanning over grayscale scanning is in storage space savings. Because each grid cell contains a value of 0 or 1, only one bit of storage space is required. Compare this with 8 bits per cell to store grayscale values of 0 through 255, and 24 bits per cell to store the red, green, and blue 0 to 255 values with a color scan. Assuming no compression, a grayscale scan will consume eight times as much disk space, and a color scan, 24 times more space than a binary threshold scan.

Scanning Density

A scanner samples an image at a user-specified density of samples, usually called dots per inch (dpi). A trade-off exists between scanning density and storage space requirements: the higher the dpi, the greater the file size. For instance, the following illustrations show the same map scanned at three different dpi levels. The binary scan at 150 dpi would typically require four times the file storage space as the 75 dpi scan of the same map. In turn, raising dpi to 300 would quadruple file size compared to the 150 dpi scan.

Binary scan of topographic map at 75 dpi.

Binary scan of
topographic map
at 150 dpi.

Binary scan of topographic
map at 300 dpi.

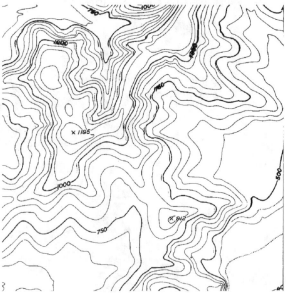

Computing Scan File Size Requirements

Scanning maps and images can quickly fill up a hard drive. Consider the following scenarios.

Assume that you are working with an old panchromatic photograph of a river floodplain. How much disk space would it take to store the photograph as an uncompressed digital image if you scanned it at 600 dpi, with photo format of 9" x 9"? Because you are scanning in grayscale, each scan cell will require 1 byte. An inch across requires 600 columns, and an inch down 600 rows; thus, a square inch would require about 360,000 bytes of disk space. To scan a 9" x 9" photo as a grayscale scan would then require (9 x 600) (9 x 600) (1 byte/cell) = about 29 million bytes.

Assume that you are working with a colored USGS 14" x 18" quad. How much disk space would you need to store this map as a digital image if it were scanned as a 600 dpi color image and stored uncompressed? Recall that in color scanning, 3 bytes are required for the three primary colors (red, green, and blue). Consequently, each scan cell requires 3 bytes. A 600 dpi square inch of color scan would require (600 columns) (600 rows) (3 bytes/cell) = 1,080,000 bytes. To scan the entire map in color would require (14 x 600) (18 x 600) (3 bytes/cell) = about 272 million bytes.

In contrast, a black and white scan would only require about 11 million bytes, because each cell in a binary threshold scan requires one bit (or 1/8 byte) of disk space.

Packaging Scanned Images

Generic Image Formats

Digital images can be packaged in three generic formats: band sequential (*.bsq*), band interleaved by line (*.bil*) and ba interleaved by pixel (*.bip*). In band sequential packaging, the first value of every pixel is stored sequentially, followed by the second value of each pixel, and so on. In band interleaved by line packaging, the first value of every image row is stored sequentially, followed by the second value of each image row, and so on. In band interleaved by pixel packaging, the values from each pixel are stored sequentially.

Proprietary Image Formats

Both ARC/INFO and ArcView can use the three generic image packaging formats. In addition, both programs can work with the proprietary image packaging formats in the next table.

Image format	File extension	Common use	Supported by ARC/INFO 7.1	Supported by ArcView 3.0
ARC Digitized Raster Graphics	.adrg	Former Defense Mapping Agency format	Yes	No
Windows bitmap images	.bmp	Microsoft Windows	Yes	Yes
ERDAS, Inc.	.gis, .lan, .img	ERDAS software	Yes	Yes
Geographical Resource Analysis Support System	.grass	GRASS GIS	Yes	No
JPEG	.jpg, .jfif	JPEG image compression	Yes	Yes, with JPEG extension
Run-length compressed	.rlc	RLC image compression	Yes	Yes
Sun raster images	.sun, .rs, .ras	Sun Openwindows	Yes	Yes
Tag image file format	.tif	Apple Macintosh, desktop publishing	Yes	Yes
ARC/INFO GRID formats		ARC/INFO GRID module	Yes	Yes, with Spatial Analyst extension

In addition, ArcView supports the following proprietary image formats for hot linked images: X-Windows Bitmap, X-Windows Dump Format (*.xwd*), MacPaint (Apple Macintosh), Graphic Interchange Format (*.gif*), and Microsoft Device Independent Bitmap (*.dib*).

Heads-up Digitizing Using ARCSCAN

Building a database is the most expensive and tedious part of a GIS project. Maps are often used for database development. You can digitize maps from a digitizing tablet or digitize scanned maps from your computer screen. Heads-up, or on-screen, digitizing from a scanned map is superior to conventional tablet digitizing of maps for two reasons. First, you can easily zoom in on an area and digitize within the enlarged region. This procedure is easier on your eyes and more accurate than tablet digitizing from paper or mylar maps. In addition, when performing tablet digitizing, accuracy is likely to suffer as you tire. With scanned map digitizing, you can use auto-trace mode to minimize variations in accuracy.

To perform map scanning and digitizing, you need a clean and clear base map. Scanning can be carried out in-house with an accurate scanner; many GIS users, however, contract scanning to a service company. The cost for scanning has dropped significantly in recent years. You can expect to pay $10 to $20 for scanning a *separate*, the size of a USGS 7.5-minute quad (approximately 18" x 24"). Service bureaus typically package images in a non-grid format.

To use ARCSCAN for heads-up digitizing, the first step is to convert a scanned map image format to an ARC/INFO grid format. ARCSCAN provides grid editing tools to remove materials that do not require conversion. ARCSCAN also offers semi-automatic and manual tracing options, as well as many parameters that affect how the tracing is done. Although ARCSCAN is considered an ARC/INFO module, it operates in the ARCTOOLS environment and can also be used in ARCEDIT for editing grids. Because the ARCSCAN module is licensed as an extension to ARC/INFO, you must determine whether you have licensed access to the module. Check the $ARCHOME/sysgen/license.dat file to view module licenses, or consult with your system administrator.

Simple Raster to Vector Data Conversion

The exercise in this section covers coversion of a scanned image to an ARC/INFO grid. The grid is then edited with ARCTOOLS to trace the soil polygon arcs from the scan to a new line coverage. The *hoytmtn.tif* file is an image of soil polygons from the Hoyt Mountain quad in northern Idaho, and has already been transformed to the UTM projection. (See Chapter 5 for a discussion of methods employed in projection conversions.) The UTM coordinates of the quad corners are located in the *hoytmtn-tics* coverage.

1. Convert the scanned image from *.tif* format to an ARC/INFO grid with the following command.

   ```
   Arc: imagegrid hoytmtn.tif hoytmtn_gd
   ```

2. Start ARCTOOLS and display the grid on the canvas.

   ```
   Arc: arctools edit
   ```

3. In the Edit Tools menu, select File | Grid Open. Choose *hoytmtn_gd* in the Select an Edit Grid menu. Click OK to display the *hoytmtn_gd* grid on the canvas and open the Grid Editing menu.

The hoytmtn_gd grid.

Grid Editing menu.

4. Create a new coverage with the tic locations corresponding to the corners of the quadrangle. In the Edit Tools menu, select File | Coverage: New.

New Coverage dialog.

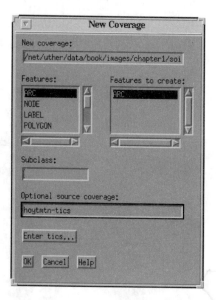

5. Enter a new coverage name of *hoytmtntrace*, and double-click on Arc in Features options to create the list. The new coverage should contain the UTM coordinates of the

quadrangle's four corners as tics. Enter *hoytmtn-tics* as the Optional Source Coverage because the existing coverage contains the tics. Then click on the OK button to create the new coverage.

6. The tics from the new coverage should be displayed at the corners of the quadrangle. Use Pan/Zoom Extent to view grid registration with the tics. Note that the soils polygons from the scan occupy only the upper portion of the quadrangle.

7. To begin autotracing arcs from the displayed grid, click on the Trace button in the Edit Arcs & Nodes menu. Place the cursor at the start of any line along the border of *hoytmtn_gd* that is well connected with other lines. A green line, called the "tracer," appears at the starting point.

✓ **TIP:** *Similar to tablet digitizing, tracing the same line twice should be avoided. A starting point along the border prevents duplicate tracing.*

Start of tracing.

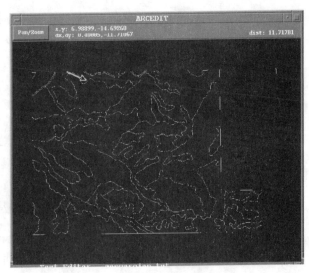

8. Press 8 on the keyboard to activate semi-automatic tracing. Lines that have been traced turn green. The green lines will become yellow after all the lines connected to the starting point have been traced, and the built-in algorithms for line straightening and generalization have been applied to the traced lines. Additional discussion of line straightening and generalization appears in the next section.

Results from semiautomatic tracing.

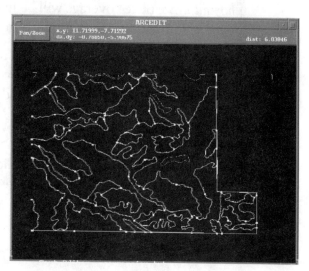

9. Lines that remain untraced in *hoytmtn_gd* are those of island polygons, polygons along the border, and tic marks. Click the Trace button in the Edit Arcs & Nodes menu, press the cursor along a untraced line, and press 8 on the keyboard.

✓ ***TIP:*** *You should change tracer direction frequently in manual tracing because you are tracing one arc at a time. Press the 2 key to change tracer direction.*

10. In the Edit Tools menu, select File | Coverage Save. If you want to build the topology for the vectorized coverage, you can select Tools | Topology Clean in the Edit Tools menu. Change the Fuzzy Tolerance to 0.00 in the Clean Polygon Topology menu, and click on Apply.

At this juncture *hoytmtn_gd* has been coverted to *hoytmtntrace*. Because the source grid was previously transformed to the UTM projection, the *hoytmtntrace* trace coverage will be in UTM coordinates. Scan transformation is discussed in Chapter 6, "Image Rectification."

Because *hoytmtn_gd* is relatively clean and clear from scanning, *hoytmtntrace* is relatively clean as well. However, there are still minor errors in *hoytmtntrace* that should be corrected in ARCEDIT.

Changing the Straightening Properties

In the Edit Arcs & Nodes menu, click the Trace Env button to open the Tracing Environment menu. Next, click on the Straighten button to open the Straighten Properties menu.

Tracing Environment menu.

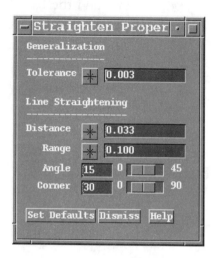

Straighten Properties menu.

Generalization and line straightening are two important properties built into ARCSCAN's tracing process. Generalization reduces the number of vertices within a line. A default value for generalization is provided (0.003" in this case). The degree of generalization is enhanced by increasing the tolerance value. Generalization can also be deactivated by setting the tolerance to zero.

The line straightening feature straightens lines at intersections and corners. Default values are initially assigned to the parameters by ARC/INFO, but can be changed. Distance, Range, and Angle values can be increased to enlarge the area affected by line straightening near an intersection. The Corner value can be increased to allow for more corners of different angles to be straightened. On the other hand, you can set Distance and Corner to zero to disable line straightening.

Changing Raster Line Parameters

A raster line refers to a line in a grid file. Upon zooming in on *hoytmtn_gd*, you will see that a raster line is comprised of an average 5 to 7 pixels. You may ask how the width of a raster line is determined. For instance, assuming that the width of a line on the base map is 0.02

inches and the base map is scanned at 300 dpi, the same line in the scanned file will have a six-pixel width (0.02 x 300). Of course, the width is not going to be the same for all raster lines in a grid file because line width is not uniform on the base map. To accommodate the varying widths of raster lines, ARCSCAN allows you to set raster line parameters.

Enlarged raster line.

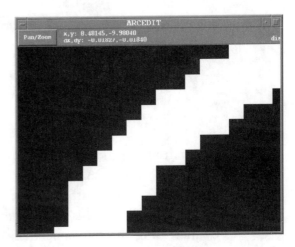

The first steps to setting raster line parameters involve accessing the Tracing Environment menu by clicking the Trace Env button in the Edit Arcs & Nodes menu. Next, click on the Line button to open the Line Properties dialog.

Line Properties dialog.

Width is a very important parameter in the Line Properties dialog. This field sets the maximum raster line width that ARCSCAN can trace. The recommended width should be about twice the average raster line width. A default value is initially assigned (0.053" in this case). As explained earlier, the average raster line width in *hoytmtn_gd* is about six pixels or 0.02 inches. Therefore, the recommended width is 0.04 inches, slightly less than the default value. Reducing the width will produce more accurate tracing; however, it may also produce more untraced lines because ARCSCAN cannot trace a line if it is wider than the specified width.

The Value parameter is used with multi-valued grids; it does not affect the tracing of *hoytmtn_gd*, a bi-level (black and white) grid file. The Gap parameter sets the maximum distance between raster line segments that ARCSCAN can jump in tracing.

The Dash parameter is used for tracing dashed lines. The Hole parameter allows the tracer to ignore a hole within the raster line that is smaller than the specified length. Dash and Hole do not affect the tracing of *hoytmtn_gd*. The Variance parameter controls how ARCSCAN reacts to variations in the raster line width. If the raster lines vary greatly in width, a decrease in variance can produce smoother traced lines.

Arc Editing Environment

The arc editing environment must be defined in an ARCEDIT session, and can also affect tracing. ARCSCAN, as set up in ARCTOOLS, functions in the ARCEDIT environment. Consequently, tracing is linked to editing tolerances for node snapping, arc snapping, and intersect arcs.

Click Edit Env in the Edit ARC & Nodes menu to open the Arc Environment Properties menu.

Arc Environment Properties menu.

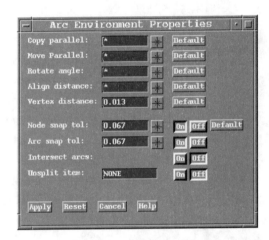

The default values for node snapping and arc snapping are assigned initially. To disable node and arc snapping for tracing arcs, you can set the tolerance values to zero.

Using ARCSCAN Utilities with ARCEDIT

Some ARC/INFO users prefer to use the ARCEDIT module rather than ARCTOOLS. You can use all ARCSCAN utilities in ARCEDIT. For instance, you can autotrace the arcs from the soils polygon scan as follows. (Before attempting the commands, verify that ARCSCAN is loaded on your system.)

```
Arc: &sys more $ARCHOME/sysgen/license.dat |grep ARCSCAN
```

Next, use ARCEDIT to trace the soil arcs from the scan grid.

```
Arcedit: display 9999 4 /**Full-screen X-Windows canvas.

Arcedit: mapextent hoytmtn_gd /***Scale canvas to fit the scan grid.

/***Specify the grid to be edited or traced from.

Arcedit: gridedit edit hoytmtn_gd

Arcedit: draw /***Draw the scan grid.

Arcedit: gridedit fillvalue 0 /***Set grid fill value to background.

/***Eliminate graphics you do not wish to trace.

Arcedit: gridedit brush many

Arcedit: gridedit save /***Save changes to the grid.

/***Create a new coverage and copy the UTM tics from hoytmtn-tics.

Arcedit: create hoytmtn-soils hoytmtn-tics

Arcedit: editfeature arc /***The focus is on arcs.

/***Specify the grid from which to trace.

Arcedit: vtrace raster hoytmtn_gd

Arcedit: vtrace gap 0 /***Do not jump gaps while tracing.

/***Begin autotrace: 2=change arrow direction, 8=autotrace save.

Arcedit: vtrace add

Arcedit: save /***Save coverage of traced soil arcs.
```

Remote Sensing

Remote sensing has been defined as "acquiring information from a distance." Remotely sensed digital images include scanned aerial photographs, radar, video, and satellite images. Selected advantages of remotely sensed images compared to traditional maps are summarized below.

- Remotely sensed data may be more current due to regular and frequent image acquisition.
- A larger variety of themes may be derived from images such as forest cover types, hydrography, water quality, agricultural cover types, and areas of eroded soils.
- Users can interpret remotely sensed data more easily. For example, traditional mapmakers portray a wetland as a pure, discrete polygon. With an image, a user can interpret wetland purity and boundaries.
- An image can cover a large area in a single snapshot. For example, a SPOT panchromatic satellite image covers approximately 360,000 ha at a 10m pixel size, while a Landsat Thematic Mapper scene covers over 3 million ha at a 30m pixel size.

The following files on the companion CD are used or referenced in this chapter.

hispat.tif	High resolution remotely sensed image of Chena River in Fairbanks, Alaska.
medspat.tif	Medium resolution remotely sensed image of Chena River in Fairbanks, Alaska.
lowspat.tif	Low resolution remotely sensed image of Chena River in Fairbanks, Alaska.
ortho.bil	Digital orthophoto from northern Idaho.
veg-polys	Vegetation polygon coverage from northern Idaho.
tmrect.bil	Landsat Thematic Mapper image of interior Alaska.
ortho.blw	World file for digital orthophoto of northern Idaho.
ortho.hdr	Header file describing digital orthophoto of northern Idaho.
ortho.stx	Statistics file for digital orthophoto of northern Idaho.
tmrect.blw	World file for Landsat Thematic Mapper image of Alaskan interior.
tmrect.hdr	Header file describing Landsat Thematic Mapper image of Alaskan interior.
tmrect.stx	Statistics file for Landsat Thematic Mapper image of Alaskan interior.
remote-sensing.doc	ARC/INFO documentation file.

Spectral Regions

Remotely sensed images are based on spectral regions from a continuum of wavelengths known as the electromagnetic spectrum. Trade-offs can be generally summarized as the shorter the wavelength, the greater the atmospheric scattering. In brief, images from the blue spectral region are relatively hazy while images from the near and mid-infrared spectral regions appear much clearer. However, the longer the wavelength, the less the energy content. Therefore, thermal images usually have a larger ground pixel size relative to images from the visible spectral regions.

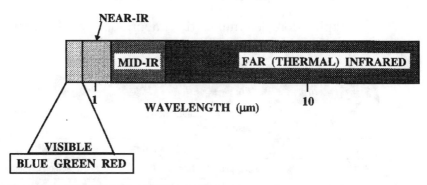

A portion of the electromagnetic spectrum.

Many spectral regions are typically used in remote sensing because surfaces in each spectral region typically have a characteristic pattern of spectral reflectance. For example, because the green reflectance of aspen forest canopy is higher than a spruce forest canopy, aspen appear as a lighter green to the human eye. You could use a digital image from the green spectral region to map aspen versus spruce stands. However, the near infrared spectral region would be even better.

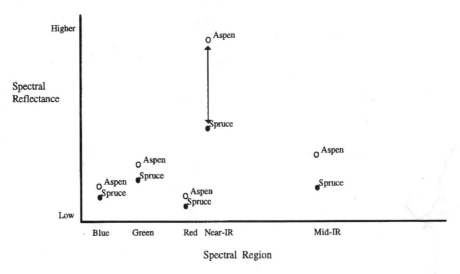

Spectral reflectance of aspen versus spruce forest. The near infrared spectral region is

the best region for separating these two forest types.

The application determines the importance of each spectral region. The table below presents selected frequently used regions.

Spectral region	Characteristics	Application
Blue	Problem with scatter, water transmittance	Mapping underwater features
Green	Low reflectance in clear water	Mapping oligotrophic waters
Red	Absorption by chlorophyll	Vegetation mapping
Mid-infrared	Absorption by water, emittance by fires	Fire and water mapping, clouds versus snow
Thermal infrared	Thermal radiance	Water/land temperatures
Microwave	"See-through clouds"	Radar remote sensing

The next illustration shows a contour map of Fairbanks, Alaska. Broadleaf vegetation typically grows on southerly-facing slopes and coniferous vegetation typically grows on northerly-facing slopes in this area.

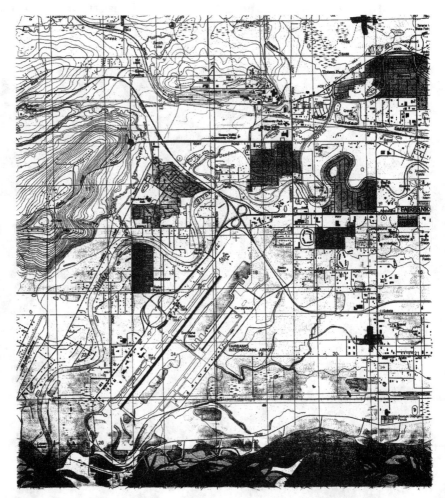

Fairbanks D2-SW quadrangle.

The next three illustrations show the same area of Fairbanks from three different spectral regions. The Fairbanks International Airport is at the lower left of the images, and all have 30m ground pixels. The large river is the Tanana River, which transports turbid glacial water and is therefore highly reflective in the red spectral region. The smaller Chena River is a clear water stream and as such has relatively low reflectance in the three spectral regions. Key characteristics of the three spectral regions are listed in the following table.

Image of Fairbanks from the red spectral region.

Image of Fairbanks from the near infrared spectral region.

Image of Fairbanks from the mid-infrared spectral regions.

Typical Surface Reflectance Responses

	Clear water	*Turbid water*	*Broadleafs*
Red reflectance	Very low	High	Low
Near IR reflectance	Very low	Low	High
Mid IR reflectance	Very low	Very low	Low

Image Resolution

Four types of resolution are used to describe remotely sensed imaging: spatial, spectral, temporal, and radiometric. All types are discussed in subsequent sections.

Spatial Resolution

Images with relatively small ground pixel sizes are referred to as high spatial resolution images. For example, SPOT PAN images are acquired with approximately 10m ground pixel size. Therefore, SPOT PAN images have high spatial resolution relative to AVHRR (advanced very high resolution radiometer) images, which are acquired with ground pixel sizes exceeding 1,000m. Photographic camera lenses are analogous to varying levels of spatial resolution in remotely sensed images. A zoom lens provides a higher spatial resolution photograph, but at a cost of covering a smaller area than a wide angle lens. Returning to remote sensed imaging, the new 1m images scheduled for 1997 will cover a swath of less than 20 km, whereas the 30m Landsat Thematic Mapper image covers a swath of over 180 km.

The remotely sensed images of the Chena River in Fairbanks, Alaska, shown in the next three illustrations are of the same file size and were taken simultaneously. The same images are available on the companion CD (*hispat.tif*, *medspat.tif*, and *lowspat.tif*). Note that as spatial resolution decreases, the area covered by the image increases.

High spatial resolution image of Chena River.

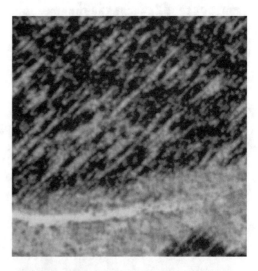

Medium spatial resolution image of Chena River.

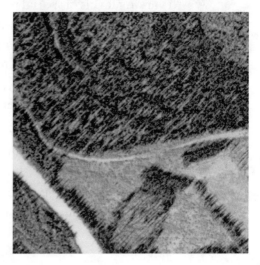

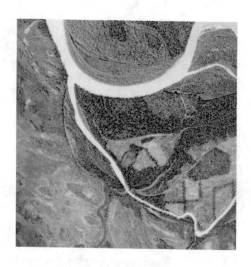

Low spatial resolution image of Chena River.

Spectral Resolution

Images from a narrow spectral region have high spectral resolution. In photography, for example, a yellow filter improves spectral resolution by eliminating short visible wavelengths and improving the sharpness of photos taken during a hazy day. In general, there is a trade-off between spectral resolution and spatial resolution. In the following two images of the Bonanza Creek Experimental Forest in interior Alaska, the lower spatial resolution image is better for delineation of spruce versus aspen because of its higher spectral resolution.

Broad spectral region (low spatial resolution) image of Bonanza Creek.

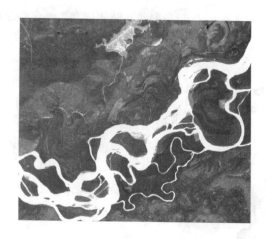

Narrow spectral region (high spectral resolution) image of Bonanza Creek.

Temporal Resolution

Images acquired on a frequent orbit cycle have high temporal resolution. A trade-off exists between spatial and temporal resolution. Generally, the more frequently data are acquired, the less detail images will have. For example, AVHRR images are acquired twice daily with ground pixel sizes exceeding 1,000m. Landsat Thematic Mapper images are acquired on a 16-day cycle with a ground pixel size of approximately 30m. Nadir images with a ground pixel size of 10m from the SPOT HRV sensor are acquired on a 26-day cycle.

Radiometric Resolution

Systems that scale spectral response variation into many values have high radiometric resolution. For example, AVHRR (advanced very high resolution radiometer) images have high radiometric resolution because the sensor scales reflect to pixel values ranging from 0 to 1023. The Landsat Multispectral Scanner (MSS) had relatively low radiometric resolution because it scaled reflectance to pixel values ranging from 0 to 63.

Satellite Imagery

Historic Data

Several different types of satellite images have been used in GIS applications. The following types of imagery can be ordered from archives. Landsat MSS images were terminated in 1992. In 1997, Landsat TM, SPOT, and AVHRR images were still being acquired on a regular orbit cycle.

Satellite Imagery Commonly Used in GIS Applications

	Landsat MSS	*Landsat TM*	*SPOT PAN/XS*	*AVHRR*
Company	USGS	EOSAT USGS	SPOT Image	NOAA USGS
World Wide Web addresses	http://edcwww.cr.wgs.gov	http://www.eosat.com	http://www.spotimage.fr	http://edcwww.cr.wgs.gov
Scene size (ha)	3.3×10^6	3.3×10^6	3.3×10^6	Huge
Cost per scene (mid-1997)	$200	$4,900 after 1986; $425 before 1986	$2,500	$190 to $32
Nominal pixel size (m) at nadir	75	30	10 (PAN), 20 (XS)	1100
Number of spectral bands	4	7	1 (PAN), 3 (XS)	5
Year of first image	1972	1982	1986	1979
Radiometric resolution	7-bit (0-127)	8-bit (0-255)	8-bit (0-255)	10-bit (0-1023)
Nadir temporal resolution	16 days	16 days	26 days	Twice daily

High Spatial Resolution Satellite Images

By the year 2000, satellite images with high spatial resolution should be available. The next table summarizes the new systems planned for 1997.

New High Spatial Resolution Satellite Imaging Systems

Satellite name	Company	World Wide Web address	Pixel size (m)
EarlyBird	EarthWatch	http://www.digitalglobe.com	3m Pan, 15m Multispectral
OrbView	OrbImage-3	http://www.orbimage.com	1m Pan, 4m Multispectral
Clark	CTA / NASA	http://www.crsp.ssc.nasa.gov/ssti/cta	3m Pan, 15m Multispectral
CRSS	SpaceImage	www.spaceimage.com	1m Pan, 4m Multispectral

Large area coverage with high resolution satellite images will not be possible. For example, EarlyBird panchromatic images will have a scene size of 3 km. Compare this to 185 km for a Landsat TM scene or 60 km covered by a SPOT scene.

Quality Control

When ordering satellite imagery, you typically have a short-term warranty for data quality. You should assess the quality of the imagery immediately after you receive it. For example, according to a pre-order computer printout, the cloud cover for the following imagery is supposed to be zero percent. By examining the mid-infrared band, you can see that there are no large clouds.

Mid-infrared image with no clouds evident.

However, a look at the blue band image of the same area reveals its weaknesses as a high-quality image. Always examine all spectral bands when checking imagery for quality.

Blue band image of same area at same time.

Sometimes poor data quality from a bad detector is not obvious. For example, the following image looks fine until you zoom in to examine individual scan lines.

Image with missing scan line.

Digital Orthophotographs

An orthophoto is a planimetrically corrected aerial photograph. This means that the effects of relief and tilt displacements on the photograph have been removed by analytical photogrammetry. Because orthos are planimetrically correct, you can use digital ortho images with a GIS.

Digital orthophotos produced by the U.S. Geological Survey (USGS) are quadrangle based, with four images per 7.5-minute quadrangle. Projected into the UTM system, these images have a pixel size of one meter. The grayscale values on digital orthophotos range from 0 to 255 (8 bits).

Digital orthophotos are also produced by service providers and GIS users. Software packages are available which can create digital orthophotos on PCs or workstations using the inputs of scanned aerial photographs, DEMs (digital elevation models), and surveyed control points. Digital orthophotos are useful to a GIS project as background coverage and as input for digitizing or updating of existing vector themes.

Digital orthophotos are typically formatted in an appropriate map projection for your GIS. They are useful as a base to check the positional accuracy of point, line, or polygon themes. A digital orthophoto called *ortho.bil* and a polygon coverage called *veg-polys* are used in exercises in the next section.

Displaying Remotely Sensed Images

Single-Band

You have already displayed gray single-band scale images from Chapter 1. Try the following exercises to drape a polygon coverage on top of a digital single-band orthophoto from northern Idaho.

Digital Orthophotos in ARCPLOT

To display a gray digital orthophoto scale image in ARCPLOT, you must first scale the canvas to fit the image. You then can use any ARC-PLOT drawing command to draw points, lines, polygons, annotations, or other features on top of the displayed image. Experiment with the following ARCPLOT commands.

```
Arcplot: display 9999 4 /***X-Windows full-screen canvas.

/***Scale canvas to fit the image called ortho.bil.

Arcplot: mapextent Image Ortho

Arcplot: image ortho.bil /***Display the image to the canvas.

Arcplot: &terminal 9999 /***Define terminal type as X-Windows.

Arcplot: lineedit /***Select a line symbol for vegetation polygons.

/***Draw the vegetation polygons on the canvas.

Arcplot: polygons veg-polys
```

ARCEDIT typically allows you to use all pan/zoom utilities, while ARCPLOT only supports the Create Window and Get Extent pan/zoom utilities. However, you can use the ARCPLOT IMAGEVIEW command for all pan/zoom utilities. Try the following commands.

```
/***Create another canvas called ortho.

Arcplot: imageview create ortho-canvas

/***Display single band image called ortho with coverage called

/***veg-polys to the canvas called ortho-canvas.

Arcplot: imageview ortho.bil 1 veg-polys ortho-canvas
```

Use the Extent or Ctrl+E utility to change the canvas map extent to a small area. Note that the vegetation polygons are poorly coregistered with the digital orthophoto. You can do a better job using ARCEDIT by on-screen digitizing with the digital orthophoto in the background.

```
Arcplot: quit /***Quit ARCPLOT and return to ARC prompt.
```

Digital Orthophotos in ARCEDIT

You can easily incorporate images in the background using the IMAGE command in ARCEDIT. In the following exercise, selected clear-cuts are digitized as a new polygon coverage.

```
Arcedit: display 9999 4 /***Full-screen X-Windows canvas.

Arcedit: mapextent image ortho.bil /***Scale canvas to fit image.

Arcedit: image ortho.bil /***Rule for displaying image.

Arcedit: backcoverage veg-polys 3 /***3 = green color.

/***Display rule for drawing polygons in background coverage.

Arcedit: backenvironment polygons

Arcedit: draw /***Display on canvas using rules.

/***On-screen digitizing of clear-cuts.

/***Create a new coverage using the tics from veg-polys coverage.

Arcedit: create clearcuts veg-polys

Arcedit: drawenvironment polygons /***Add polygons to drawing rules.

Arcedit: build /***Build initial polygon topology.

Arcedit: editfeature poly; add /***Digitize clear-cut polygons.

Arcedit: save /***Save clear-cuts coverage.

Arcedit: quit /***Quit ARCEDIT and return to ARC prompt.
```

Digital Orthophotos in ArcView

Now try displaying the digital *ortho.bil* and vegetation polygons in Arc-View.

1. Start ArcView. Choose Views in the project window, if it is not already highlighted. Click New to create a new view.

2. Select Add Theme from the View menu.

3. Choose Image as the Data Source Type. Double-click on *ortho.bil* to add it as a theme.

4. Double-click *ortho.bil* in the legend to open the Image Legend Editor.

Image Legend Editor.

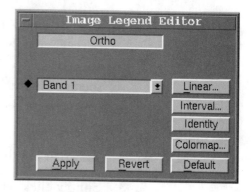

5. Click on Default to apply a 95 percent linear contrast stretch.

6. Draw the vegetation polygons on top of the image, and then select Add Theme from the View menu. Choose Feature Data Source as the Data Source Type. Double-click *veg-polys* to add it as a theme. Finally, click the check box next to *veg-polys* in the legend to superimpose *veg-polys* on the image.

7. Try changing the color of the vegetation polygon outlines. Double-click *veg-polys* in the legend to open the Legend Editor. Select Outline and a color other than black in the Legend Editor. Click on Apply to execute the new drawing rule.

8. You can use the magnifying glass icon to view the discrepancies between the *veg-polys* polygon boundaries and clearcut areas on *ortho.bil*.

↝ **NOTE:** *Finding clear-cuts on an orthophoto is easy because of their light color and regular boundaries.*

You can create a more accurate polygon theme by digitizing from the *ortho.bil* image. Try creating a new polygon theme called *clearcuts* as follows.

1. Create a new theme called *clearcuts.shp*. Select View | New Theme. In the Feature Type menu, select Polygon and click on OK. Input *clearcut.shp* as the name of the new theme. The new theme will automatically be added to the view in "start editing" mode.

2. To digitize a clear-cut polygon, select the Polygon tool, shown at left, from the Drawing tool bar.

3. Simply click with the mouse for the location of each polygon vertex. When you wish to close a polygon, double-click the mouse. When you have completed the clear-cut digitizing, select Theme | Stop Editing.

Multi-Band Images

You can simulate a color photograph if the digital image has pixel values from the blue, green, and red spectral regions. For example, assume you are working with a Landsat Thematic Mapper image. In the next table, the bands assigned to the red, green, and blue display to simulate a color photo would be bands 3, 2, and 1, respectively.

Spectral Bands for Landsat Thematic Mapper Image

Band number	Spectral region
1	blue
2	green
3	red
4	near-IR
5	mid-IR-I
6	thermal
7	mid-IR-II

You can also simulate color infrared photography with many remotely sensed images such as Landsat MSS, Landsat TM, and SPOT HRV XS imagery. Interpretation of vegetation types is generally easier in color infrared imagery because reflectance of near-infrared is greater for plants compared to visible spectral regions. In addition, because color infrared imagery does not use the short visible spectral region (blue), the images are less hazy when compared to true color images.

Blue Light	Green Light	Red Light	Near-IR Light
↓	↓	↓	↓
Filtered Out			
Black Print	Blue Print	Green Print	Red Print

Color infrared film responses.

Plants reflect highly in the near-infrared spectral region and therefore usually appear red on color infrared photography. Clear water reflection peaks in the blue spectral region and therefore appears black on color infrared photography. Muddy water reflects highly in the green and red spectral regions and therefore appears cyan on color infrared photography.

Assume that you are working with a SPOT XS image. Based on the next table, the bands you would assign to the red, green, blue display to simulate a color infrared photo are bands 3, 2, and 1, respectively. Note that the same spectral regions are represented by bands 4, 3, and 2 from the Landsat Thematic Mapper.

Spectral Bands for SPOT XS Image

Band number	Spectral region
1	green
2	red
3	near-infrared

Multi-Band Images Using ARCPLOT

In Chapter 5, you will develop a land cover theme of an area in interior Alaska. In this section the area will be displayed in true color. The *tmrect.bil* image on the companion CD is comprised of the first five bands from the Landsat Thematic Mapper. The third value in each pixel represents the red spectral region, while the second and first values represent the green and blue spectral regions, respectively. Therefore, to produce a true color image, assign the red spectral region to control the red screen intensity, the green spectral region to control the green screen intensity and the blue spectral region to control the blue screen intensity. Try this with ARCPLOT.

```
Arcplot: display 9999 4 /***ARCPLOT X-Windows full-screen canvas.

Arcplot: mapextent image tmrect.bil /***Scale canvas to fit the im-
age called tmrect.bil.

Arcplot: image tmrect.bil composite 3 2 1 /***Assign bands 3, 2, 1
to red, green, blue screen intensities, respectively.
```

Alternately, you could use the similar composite 3 2 1 flag with the IMAGEVIEW command in ARCPLOT and the IMAGE command in ARCEDIT.

To produce a color infrared image of the same area, assign the near infrared, red spectral, and green spectral regions to control the red, green, and blue screen intensities, respectively. Try using IMAGE-VIEW to display three different band combinations.

```
/***Create three imageview canvases and use the mouse to see all
three canvases.

/***Create an imageview canvas for a true color display.

Arcplot: imageview create true-color

/***Create an imageview canvas for a color infrared display.

Arcplot: imageview create color-ir

/***Use red and invisible spectral regions with this canvas.

Arcplot: imageview create funky-color

/***Simulate color photograph.

Arcplot: imageview tmrect.bil composite 3 2 1 # true-color

/***Simulate color infrared photograph.

Arcplot: imageview tmrect.bil composite 4 3 2 # color-ir

/***Use mid-ir, near-ir, red spectral regions.

Arcplot: imageview tmrect.bil composite 5 4 3 # funky-color
```

Multi-Band Images Using ArcView

1. Select Add Theme from the View menu. Double-click on *tmrect.bil* to add it as a theme.

↠ **NOTE:** *Verify selection of Data Source Type as Image Data Source.*

2. Click the check box next to the theme in the legend to display the image. Double-click on the image legend to open the Legend Editor.

3. You can now specify the image pixel values to control red, green, and blue. Try bands 3, 2, and 1 to control the red, green, and blue, respectively, and then select the Default button. Perform a 95 percent linear contrast stretch for each band. You should now see a color image.

Band assignments to display a color image.

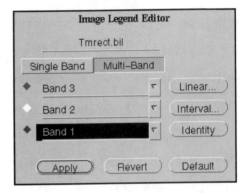

4. Try band 4 to control red, band 3 to control green, and band 2 to control blue and then select the Default button. You should now see a color infrared image. The bright red vegetation is broadleaf forest whereas the darker areas are mostly muskeg. Note how much easier it is to see these differences on the color infrared image compared to the colored image. The river is mostly glacial water that is transporting a high sediment load (known as "glacial flour").

Part III
Unwarping Images

5

Map Projection and Coordinate Systems

Images can be useless unless they match the map projection and coordinate system your GIS is using. The process of unwarping or transforming an image and assigning map coordinates to each pixel is sometimes called "rectification." Because image rectification requires an understanding of map projections and coordinate systems, this chapter provides an overview of both. The *usa* coverage and *map-project.doc* from the companion CD are used in exercises or referenced in this chapter.

Map Projections

Map projection is the process of transforming the spatial relationship of features on the Earth's surface into a flat map. A flat map, however, does not accurately reflect the shape of the Earth. For this reason, hundreds of different map projections have been used in mapmaking because the transformation from an approximation of a sphere to a flat surface always involves distortion. However, every map projection preserves certain spatial properties while sacrificing others.

You have probably encountered map projections with names like the Lambert conic conformal or Albers conic equal-area. Lambert and Albers are the names of mapmakers who originally proposed the projections. The other parts of the name describe the map projection's preserved property and projection surface. The following four classes of map projections are named according to preserved properties.

- *Conformal* preserves local shapes.
- *Equal area* (or equivalent) represents areas in correct relative size.
- *Equidistant* maintains consistency of scale for certain distances.
- *Azimuthal* (or true direction) retains certain accurate directions.

Conformal and equivalent properties are mutually exclusive, but a map projection can have more than one of the other preserved properties (e.g., conformal and azimuthal). Conformal and equivalent properties are global properties and therefore apply to the entire map projection. Equidistant and azimuthal properties are local properties and are usually accurate only from or to the center of the map projection.

Mapmakers often use a geometric object to illustrate how a map projection can be constructed. For example, by placing a cylinder tangent to a lighted globe, you can make a projection by tracing the lines of longitude and latitude onto the cylinder. The cylinder in this case is the projection surface. Other common projection surfaces include a cone and a plane. Map projections are cylindrical, conic, or azimuthal, according to whether they are constructed using a cylinder, cone, or plane, respectively.

The use of a geometric object helps explain case and aspect in map projections. Take the example of a conic projection: the cone can be placed so that it is either tangent to or intersecting the globe. The first is the *simple* type of conic projection and results in a single line of tangency, and the second is the *secant* type which results in two lines of tangency. The line of tangency has no projection distortion and its scale is called the *principal scale*. The degree of distortion increases away from the line(s) of tangency.

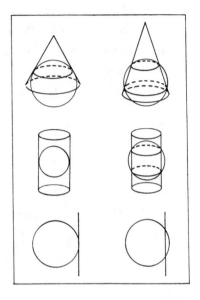

*Simple case (left),
secant case (right); and
conic, cylindrical, and
azimuthal projection
(top to bottom).*

Aspect describes the placement of a geometric object relative to a globe. A plane, for example, may be tangent at any point on a globe. Polar aspect refers to tangency at the pole; equatorial aspect, at the equator; and oblique aspect, between the equator and the pole. The degree of projection distortion increases away from the contact point between the plane and the globe.

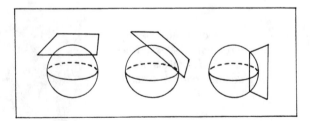

*Left to right: polar,
oblique, and
equatorial aspect.*

When you work with map projections in ARC/INFO or ArcView, you often have to define the standard parallel or the standard meridian. Both refer to the line of tangency or the line with no distortion. Again, take the example of a conic projection. A simple conic projection has a single standard parallel, whereas a secant conic projection has two standard parallels (usually called the first and second standard parallels). The degree of distortion increases the greater the distance from the standard parallel(s). The following illustrations were produced using the same GIS theme. Note how distortion decreases as the standard parallels for the map projection are changed.

Albers conic equal-area projection of Alaska with standard parallels at 30 and 45 degrees south.

Albers conic equal-area projection of Alaska with standard parallels at 20 degrees north and south.

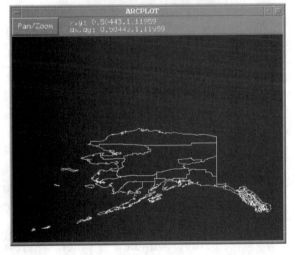

Albers conic equa-area projection of Alaska with standard parallels at 55 and 65 degrees north.

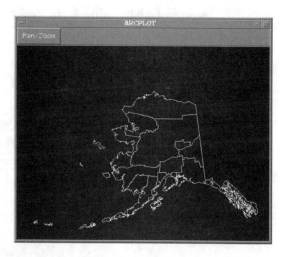

↝ **NOTE:** *The standard parallel is often different from the central parallel, which is the central line of a map projection. Likewise, the standard meridian may be different from the central meridian. The central parallel and meridian define the center of a map projection.*

To assist users in choosing from hundreds of map projections, cartographers sometimes group them by mapping application area: the world, a hemisphere, a continent, a country, or a region. This is the approach used in the ArcView GIS software's predefined map projections.

Commonly Used Map Projections

Mercator

The Mercator is probably the most well-known projection in use. This projection was invented by a Flemish cartographer in the 1500s for navigational purposes. Maps with this projection show true courses as straight lines. The Mercator is a conformal projection, and the equator is typically the standard parallel. This projection requires the following parameters: latitude of true scale (standard parallel), longitude of central meridian, false easting (x value at the central meridian), and false northing (y value at the standard parallel).

Mercator projection with equator as standard parallel. Note that the map scale increases in proportion to distance north or south of the equator.

Mercator projection with equator as standard parallel. Note that the scale factor increases dramatically as you go north from the standard parallel.

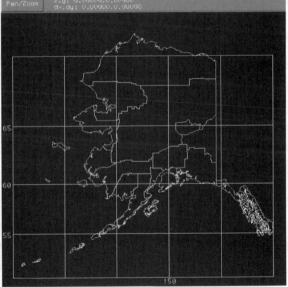

Transverse Mercator

The transverse Mercator is used for two common coordinate systems: the Universal Transverse Mercator (UTM) grid system, and the State Plane Coordinate (SPC) system in states with north-south orientation SPC zones. ARC/INFO and ArcView require the following parameters for the projection: longitude and scale factor of central meridian, false easting or x-value along the central meridian, latitude of origin (or central parallel), and false northing or y-value along the central parallel. False easting and northing are the x-shift and y-shift values assigned relative to the center of the projection.

For a UTM zone, the central meridian is assigned a false northing of 500,000m and the central meridian [**central meridian for both?**] is assigned a scale factor of 0.9996. With this scale factor, parallel meridians lying 180 km east and west of the central meridian have a scale factor of 1. The equator is the latitude of origin for the UTM system and is assigned a northing of zero. In the next illustration, the central meridian has a scale factor of 1.000. Note that as you move east or west from the central meridian, the local map scale becomes larger.

Transverse Mercator projection with -150 degrees as central meridian.

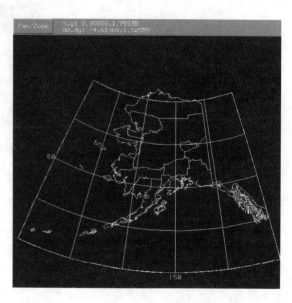

Albers Conic Equal-Area

The Albers conic equal-area projection is a good choice for a mid-latitude area of greater east-west than north-south extent, such as the coterminous United States. It is typically used as a secant projection. The parameters for this projection include first and second standard parallels, central meridian, latitude of projection's origin, false easting (or x-value) of the central meridian, and false northing (or y-value) of the projection's origin. The scale factor along the two standard parallels is 1.0000. Scale is reduced between the two standard parallels and increased north and south of the two standard parallels.

Albers conic equal-area projection of Alaska with standard parallels at 55 and 65 degrees, central parallel at 60 degrees, and central meridian at -150 degrees.

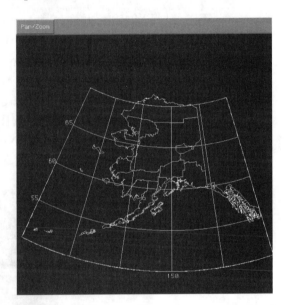

Lambert Conic Conformal

The Lambert conic conformal projection is used for the state plane coordinate system in states that spread east-west. It looks similar to the Albers conic equal-area projection, but it is conformal rather than equal area. This projection requires the same parameters as the Albers conic equal-area projection.

Polyconic

The polyconic projection was used until the mid-1950s for standard topographic maps of the United States. The projection is comprised of a series of conic projections stacked together; therefore, it has many standard parallels. Unlike other conic projections, the polyconic projection has curved rather than straight meridians. Consequently, it would be a poor projection for tiles across large areas. The parameters required for the polyconic projection are longitude of central meridian, latitude of projection's origin, false easting of the central meridian, and false northing of the projection's origin.

Datums

The Earth only approximates a sphere. For small-scale maps such as atlases, mapmakers treat the earth as a sphere. But for large-scale maps (1:1,000,000 or larger), the Earth must be treated as an *ellipsoid, oblate ellipsoid,* or *spheroid* (an ellipsoid that approximates a sphere). Because an ellipsoid is based on an ellipse, the two measures used to describe an ellipsoid are the semi-major axis and semi-minor axis. In the United States, Clarke 1866 was the standard spheroid for mapping until the mid-1980s. The semi-major and semi-minor axes for Clarke 1866 measure 6,378,206.4m and 6,356,583.8m, respectively. The difference between the axes, although small on the global scale, has an impact on large-scale mapping.

> ➔ **NOTE:** *Ellipsoid and spheroid are used interchangeably in ARC/INFO and cartography textbooks.*

Clarke 1866 represents a ground-measured spheroid. Ground-measured spheroids, however, are being replaced by satellite-determined spheroids. WGS (World Geodetic System) 84 is a spheroid determined from satellite orbital data; its semi-major axis measures 6,378,137m and semi-minor axis, 6,356,752.3m. As satellite technology continues to improve, more accurate measurements of the Earth will continue to be incorporated into new spheroids.

> *NOTE:* WGS 84 *is considered the same as GRS 80 (Geodetic Reference System 1980) for mapping concerns because both have nearly the same semi-major axis and semi-minor axis values.*

Spheroids are used as the bases for horizontal control systems. NAD 27 (North American Datum of 1927) is based on the Clarke 1866 spheroid, and NAD 83 (North American Datum of 1983) is based on the GRS 80 or WGS 84 spheroid. As you may have observed on some USGS 7.5-minute quads, the locational shift between NAD 27 and NAD 83 can be substantial. For example, for the Moscow East quad in Idaho (117 W, 46.75 N), the shift is 78m to the east and 16m to the north, while for the Fairbanks area (148.0 W, 64.75 N) in Alaska, the shift is 115m east and 47m north.

> *NOTE 1: The default datum in ARC/INFO is NAD27. Because many GIS professionals are converting from NAD27 to NAD83, you are advised to verify the datum of a coverage before using it. Use the ARC/INFO DESCRIBE command to obtain information on datum and spheroid.*

> *NOTE 2: Vertical control is as important as horizontal control. Elevations are made relative to a spheroid and a vertical datum. The North American Vertical Datum of 1988 (NAVD 88) is the basis for vertical control networks in the United States.*

Coordinate Systems

The next time you pick up a USGS 7.5-minute quad, read the map margin and legend. In addition to the map projection used, the legend describes the coordinate systems available from the map margin. The three most common coordinate systems are geographic, UTM, and State Plane.

Longitude and latitude pertain to the geographic coordinate system, a spherical coordinate system. The equator is 0 degree latitude, and the prime meridian, 0 degree longitude; both serve as the base lines for geographic coordinates. You have probably used map coverages measured in longitude and latitude values because a large proportion of the data available from commercial vendors is stored in geographic coordinates.

Each USGS 7.5-minute quad has eight point locations with precise longitude and latitude readings, which can be used as tic locations in digitizing. Four are at the corners of the quad, and the other four are inside the quad denoted by black plus signs. The four corners cover the extent of 7.5 minutes in longitude and latitude, and the inner four represent the division in 2.5 minutes.

One problem with spherical coordinates is that the calculation of area, perimeter, and length is complicated compared to calculations in planar coordinates. You can develop a plane coordinate system by placing a rectangular grid on a map projection. Because they are Cartesian coordinates, plane coordinates have x and y values (easting and northing).

Plane coordinate systems are typically used on large-scale maps. The distortions associated with map projections are minimal on large-scale maps; therefore, detailed calculations and positioning can be achieved using plane coordinates. Three coordinate systems commonly used in the United States are the Universal Transverse Mercator (UTM), Universal Polar Stereographic (UPS), and State Plane Coordinate (SPC).

Universal Transverse Mercator

The UTM was developed by the U.S. Army in 1947 for designating coordinates on large-scale maps worldwide. This grid system divides the Earth between 84 degrees N and 80 degrees S into 60 zones. Each zone is 6 degrees longitude wide with zone 1 beginning at 180 to 174 degrees West. Each of the 60 zones is mapped onto the transverse Mercator projection, with a scale factor of 0.9996 at the central meridian. A false origin is then assigned to each UTM zone.

In the United States, UTM coordinates are measured from a false origin located 500,000m west of the central meridian of the zone and at the equator. For example, a false easting of 640,000 and a northing of 7,181,000 represents a point 140,000m east of the zone's central meridian and 7,181,000m north of the equator. UTM tics are typically printed at 1,000m intervals along the map margin of USGS 7.5-minute quadrangles.

The initial station for the North American Datum of 1927 (Meades Ranch, Kansas) is located at 98 degrees 32 minutes 30.506 seconds W longitude and 39 degrees 13 minutes 26.686 seconds N latitude. Each UTM zone is six degrees wide, with zone 1 starting at 180 degrees W. Zone 1 ranges from 180 to 174 degrees W; Zone 2 from 174 to 168 degrees W, and so on. The zone that encompasses 98 degrees W is UTM zone 14, 102 to 96 W.

Universal Polar Stereographic

The Universal Polar Stereographic (UPS) covers the polar areas. The UPS is centered on the pole and is used for dividing the polar area into a series of 100,000m squares similar to the UTM grid system.

State Plane Coordinate

The State Plane Coordinate (SPC) was developed in the 1930s by the U.S. Coast and Geodetic Survey so that each of the 50 states could permanently record original land survey monument locations. The SPC system provides a common survey datum for referencing horizontal control of surveys over a relatively large area. To maintain the required accuracy of one part in 10,000 or less, the width of a zone is held to a maximum of 158 miles; consequently, there may be several SPC zones within a state. The system is usually based on the transverse Mercator projection for north-south zones and the Lambert conformal conic projection for east-west zones. Point locations within each SPC zone are typically measured in feet from a false origin. However, since the adoption of the NAD83, new computations and grids are specified in meters.

National Map Accuracy Standards

National standards for the horizontal and vertical accuracy of USGS topographic maps, revised and adopted in 1947, are still in effect. The standards for horizontal accuracy require that no more than 10 percent of the well-defined map points tested shall be more than one-thirtieth of an inch at scales larger than 1:20,000, and one-fiftieth of an inch at smaller scales. This means that the tolerance is 40 feet on the ground for 1:24,000 scale maps, and about 100 feet on the ground for 1:62,500 scale maps. The standards for vertical accuracy require that no more than 10 percent of the elevations tested shall be in error more than one-half the contour interval for maps on all publication scales.

Map Projections and Coordinate Systems in ARC/INFO

ARC/INFO supports 26 spheroids and 46 projections or coordinate systems. The PROJECT command allows you to convert a coverage, grid, or file between two projections or coordinate systems, including the geographical coordinate system. The choice of a spheroid is specified with the SPHEROID subcommand, and the choice of a projection or coordinate system is made with the PROJECTION subcommand.

⊶ ***NOTE:*** *ARC/INFO combines map projections with coordinate systems.*

In the following, a coverage called *usa* is projected from geographic coordinates to the Albers conic equal-area projection.

```
Arc: project cover usa usaalbers /***Project the usa coverage to
usaalbers.

Project: output /***The input projection information is available;
start with the output.

Project: projection albers /***Albers is the output projection.

Project: units meters /***Measurement unit in meters.

Project: parameters /***Begin the parameters.

Project: 33 00 00 /***33 degrees north is the first standard parallel.

Project: 45 00 00 /***45 degrees north is the second standard parallel.

Project: -120 00 00 /***120 degrees west is the central meridian.

Project: 42 00 00 /***42 degrees north is the central parallel.

Project: 0 /***False easting is zero.

Project: 0 /***False northing is zero.

Project: end /***End of the PROJECT command.
```

Next, plot the projected coverage called *usaalbers*, and use the ARC-PLOT commands NEATLINE and NEATLINEGRID to add lines of longitude and latitude as a frame.

```
Arcplot: display 9999 4 /***Full-screen X-terminal canvas.

/***Scale canvas to fit extent of usaalbers.

Arcplot: mapextent usaalbers

Arcplot: arcs usaalbers /***Draw arcs from usaalbers.

/***Specify prj.adf as the input projection and prj.adf as output.

Arcplot: mapprojection ./usa/prj.adf ./usaalbers/prj.adf

/***Plot long/lat outside the usaalbers map extent.
```

```
Arcplot: clipmapextent off

Arcplot: linecolor green /***Specify green as the line color.

Arcplot: neatline -180 20 -60 70 ./usa/prj.adf /***Specify the neat-
line extending from 180 to 60 W and 20 to 70 N.

Arcplot: neatlinegrid 30 10 ./usa/prj.adf /***Specify 30 degrees as
E-W interval and 10 degrees N-S interval.
```

> ⚬➤ **NOTE 1:** *Once the coverage projection is defined,* prj.adf *is the projection file that resides in the coverage directory.*

> ⚬➤ **NOTE 2:** *The* MAPPROJECTION *command changes the coordinates for plotting to those specified in the input projection file, such as the geographic coordinates in the above example. To change the coordinates back to the Albers projection, type the following:* Mapprojection Off.

Map Projections and Coordinate Systems in ArcView

Similar to ARC/INFO, ArcView can also handle a large number of map projections and coordinate systems. You can choose from predefined projections or coordinate systems in ArcView, or select a map projection and define your own parameter values. The predefined systems cover the categories listed below.

- Projections of the world
- Projections of a hemisphere
- Projections of the United States
- State Plane - 1927
- State Plane - 1983
- UTM
- National grids

Each category above has a select list. For example, projections of the United States include an Albers equal-area projection for Alaska, coterminous United States, Hawaii, or North America; an equidistant conic projection for the coterminous United States or North America; and a Lambert conformal conic projection for the coterminous United States or North America. Every projection is predefined with respective ellipsoid and parameter values.

You can also choose a custom projection. ArcView offers 20 projections, and 12 ellipsoids. After you select a system, you can then define the parameter values required for the system.

As seen in the following steps, projection from geographic coordinates in decimal degrees to real world coordinates in ArcView is easy.

1.   Select View | Add Theme. Double-click on the *usa* coverage to add it as a theme.

2.   Click the check box next to the theme (*Cnty*) in the legend to display *usa*.

The usa coverage in geographic coordinates.

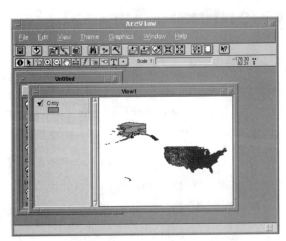

3. Select View | Properties to access the View Properties dialog.

View Properties dialog.

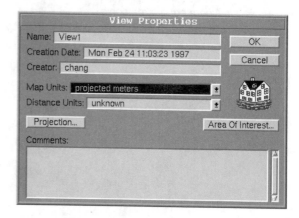

4. Select projected meters for the Map Units. Click on the Projection button to open the Projection Properties dialog box.

Projection Properties dialog.

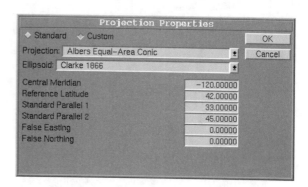

5. Click the button next to Custom, and select Albers equal-area conic for the projection. Change Central Meridian to -120.0, Reference Latitude to 42.0, Standard Parallel 1 to 33.0, and Standard Parallel 2 to 45.0. Click OK.

The usa coverage in the Albers equal-area conic projection.

Image Rectification

It is possible to unwarp images with simple linear models as long as the images do not have significant tilt and relief displacement. Analytical photogrammetric methods are required to unwarp aerial photographs of mountainous areas because of such displacements. However, aerial photographs of relatively flat landscapes, satellite images acquired at nadir, or scanned maps can often be transformed for GIS applications using linear models known as *affine transformations*.

In this chapter, a SPOT panchromatic satellite image is rectified to the UTM projection. In addition, two scanned maps are transformed from scanner pixel coordinates to the UTM projection. The following files from the companion CD are required to perform exercises in this chapter.

gps-rds	Line coverage of logging roads acquired with a GPS receiver.
spot-pan.bil	10m SPOT panchromatic satellite image to be transformed using gps-rds.
scanquad.bil	Scanned map covering the same area as spotpan.bil and gpsrds.
npole.tif	Scanned topographic map of North Pole, Alaska, for the ArcView grid rectification exercise.
rectifyscript.txt	Avenue script to rectify images.
spot-pan.hdr	Header file describing SPOT HRV image.
spot-pan.stx	Statistics file for SPOT HRV image.

scanquad.hdr	Header file describing Fairbanks, Alaska, D2 quad image.
scanquad.stx	Statistics file for Fairbanks, Alaska, D2 quad image.
longlat-tics.e00	Tic coverage in geographic coordinates.
utm-tics.e00	Tic coverage in UTM coordinates.
rectification.doc	ARC/INFO documentation file.

Affine Transformation

You have probably used the affine transformation in everyday GIS work. When you register a map on a digitizing tablet, you typically use an affine transformation to transform digitizer coordinates to real world coordinates. Whenever you use the ARC/INFO TRANSFORM command, you are most likely building an affine transformation to transform from local to real world coordinates.

In ARC/INFO and ArcView, the affine transformation model coefficients are stored in a special companion file. This special file is called the "world file" because it allows the GIS to translate from pixel coordinates to real world coordinates. For example, the world file for a *.bil* packaged image would have a *.blw* extension and would contain the affine transformation coefficients.

Assume that you want to transform an image to fit the UTM grid system. The basic process requires that you identify image pixels which correspond to the ground control points in UTM coordinates. As seen below, the affine transformation essentially consists of two linear equations for a plane.

```
Xutm = a +b1(image column number) + b2(image row number)

Yutm = a + b3(image row number) + b4(image column number)
```

Another way to express these equations is as follows.

```
Xutm = a +b1(Xpixel) + b2(Ypixel)

Yutm = a +b3(Ypixel) + b4(Xpixel)
```

Developing Affine Transformation Models: A Simple Example

This section presents a simple theoretical example of developing affine transformation equations. In the example, the coordinates of four locations known as *ground control points* (GCPs) have been estimated from a map or by using a GPS. The next step is to display an image of the area and locate the image pixels for the four ground control locations. Typically, ground control locations are road intersections, rock outcroppings, or the center point of a small pond, features which ideally show up clearly as a single distinct grid cell on an image. A link is a ground control location that has both real world (UTM in the example) and image coordinates.

Image coordinates			Real world coordinates	
GCP	Xpixel	Ypixel	Xutm	Yutm
1	200	1700	429 000	7 183 000
2	400	1600	431 000	7 184 000
3	100	1900	428 000	7 181 000
4	500	2000	432 000	7 180 000

The relationships between pixel and map coordinates can be plotted as shown in the next illustration.

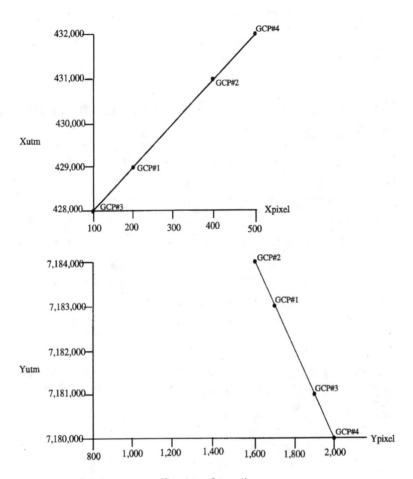

Linear models known as affine transformation.

Using the data from the above example, the affine transformation equations are calculated as shown below.

```
Xutm = 427,000 + 10(Xpixel) + 0(Ypixel)

Yutm = 7,200,000 - 10(Ypixel) + 0(Xpixel)
```

Image Adjustments By Affine Transformation

The affine transformation can be used to unwarp images to match a map projection and coordinate system by making adjustments for scale, translation, rotation, and skewness.

Scale

The affine transformation changes image scale by expanding or reducing in the x and/or y directions. This section presents a simple example that illustrates scaling transformation. Assume that you have identified four pixels in an image and that their map coordinates have been estimated as shown in the following table.

GCP#	Image column	Image row	Map X	Map Y
1	125	500	2500	5000
2	250	500	5000	5000
3	125	1000	2500	2500
4	250	1000	5000	2500

The affine transformation model follows:

```
Xmap = 0.0 + 20(Xpixel) + 0.0(Ypixel)

Ymap = 7500 - 5(Ypixel) + 0.0(Xpixel)
```

The 20(Xpixel) coefficient stretches the image pixels by a map unit of 20 in the x direction, while the 5(Ypixel) coefficient stretches the image pixels by a map unit of 5 in the y direction.

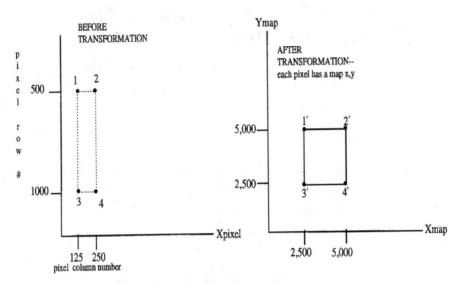

Change in scale from the affine transformation.

Translation

For a translation adjustment, the x and/or y origin is shifted. A simple example appears below.

GCP#	Image column	Image row	Map X'	Map Y'
1	7,000	6,000	4,000	4,000
2	10,000	6,000	7,000	4,000
3	7,000	3,000	4,000	1,000
4	10,000	3,000	7,000	1,000

The affine transformation equations for the translation are as follows.

```
Xmap = -3,000 + 1.0(Xpixel) + 0.0(Ypixel)

Ymap = -2,000 +1.0(Ypixel) + 0.0(Xpixel)
```

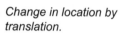

Change in location by translation.

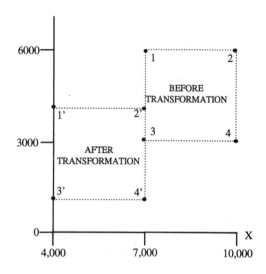

Rotation

Many polar orbiting satellites continuously image the Earth's surface at mid-morning. Because the Earth is rotating and the satellite is imaging the Earth at the same time it orbits, the satellite orbit is from northeast to southwest. Therefore, satellite images are oriented with northeast rather than true north at the top of the image. An affine transformation is typically used to rotate the image so that it is correctly oriented with north at the top and south at the bottom of the image. In a rotation adjustment, the x and y axes are rotated from the origin as demonstrated in the following example.

GCP#	Image column	Image row	Map X'	Map Y'
1	0	625	250	500
2	250	750	500	500
3	125	375	250	250
4	375	500	500	250

The affine translation equations for the rotation are as follows:

$$Xmap = 0.8(Xpixel) + 0.4(Ypixel)$$

$$Ymap = 0.8(Ypixel) + -0.4(Xpixel)$$

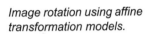

Image rotation using affine transformation models.

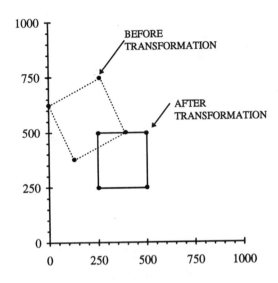

Unwarping Images in ARC/INFO

The REGISTER command in ARC/INFO initiates an interactive program that allows you to develop an affine transformation for rectifying images. Using this program, point to each image GCP pixel and then enter the associated map coordinate for that pixel by either typing in the map coordinates, or by pointing to the same location on a coverage that contains the map coordinates. In the following example, the *gps-rds* coverage was developed by driving the logging roads with a GPS receiver, applying differential corrections to the GPS estimates, and then generating the coverage in UTM coordinates.

In this exercise, the REGISTER program is used to develop affine transformations and unwarp the *spot-pan.bil* image.

> ✓ **TIP:** *Before starting the REGISTER program, verify that ARC-DOC has been started; otherwise, you may have difficulty seeing the frames in the IMAGE or COVERAGE windows.*

First, enter the following command line at the Arc: prompt.

```
/***gps-rds arcs will be colored red, result of using 2.

Arc: register spotpan.bil gps-rds 2
```

The REGISTER program will open several windows on your terminal.

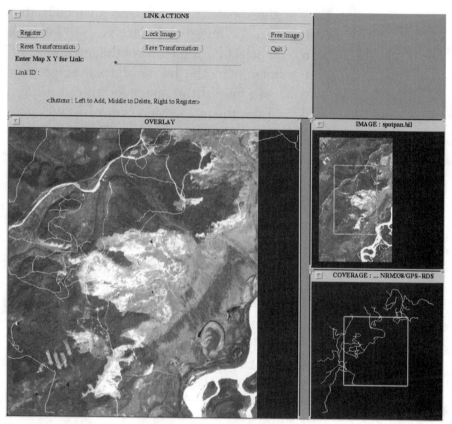

Graphic windows associated with the REGISTER program.

Note that there is a frame in the Image and Coverage windows. Use the Select button on the mouse to move the frames and the Adjust button on the mouse to adjust the size of each frame in the Image and Coverage windows. When you press the Menu button on the mouse, the areas inside the Image and Coverage frames are displayed in the Overlay window.

In the following exercise, you will quickly create four crude links and use them to develop an initial affine transformation. The initial transformation can be used to temporarily transform or "lock" the image. Once the image has been temporarily locked, the roads will line up much closer to the image road locations.

1. Use the mouse Adjust button to reduce the size of the frame in the Image and Coverage windows. Next, use the mouse Select button to move the frame over the same road intersection in the Image and Coverage windows. Once you have framed the same road intersection in both windows, press the mouse Menu key while you are in either the Image and Coverage window. The road intersection from both these windows will display in the Overlay window.

2. In the Overlay window, use the mouse to select a road intersection on the image, and then select the same location on the roads coverage. The image and UTM coordinates of the link will be displayed in the Link Actions window. Repeat the process, selecting other intersections, until you have developed four links. The next illustration shows the initial four links developed by selecting road locations from the image and the associated *gps-roads* coverage. Link 1 is a sharp bend in the road on the image and the *gps-roads* coverage. Links 2 through 4 are road intersections that are evident in both the image and the *gps-roads* coverage.

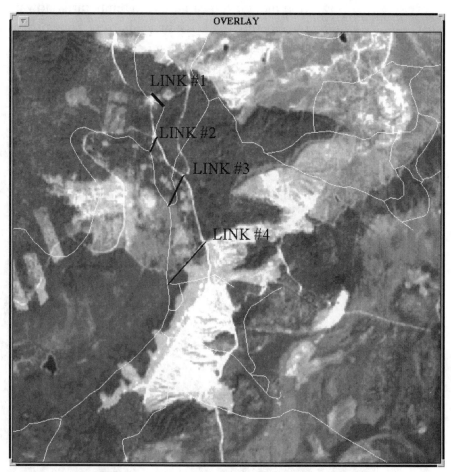

Four crude links developed for the initial affine transformation.

3. Once you have developed four crude links, you can use them to build an initial transformation and temporarily unwarp the image. To develop the initial affine transformation, click on the Register button. A Registration Results window will be displayed. Because you are using only four links, ignore this window and select the Done button to dismiss it. Once you develop an initial affine transformation model, you can use it to temporarily unwarp the image by selecting the Lock Image button. Selection of additional links will then become much easier.

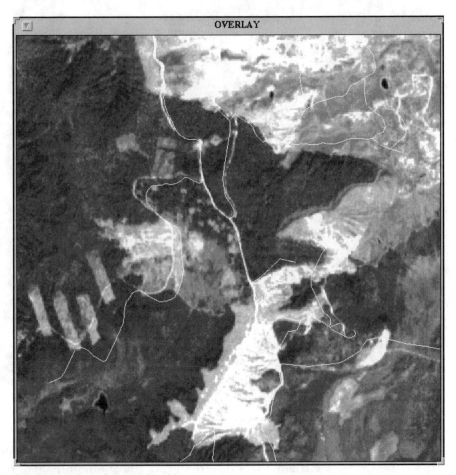

Overlay window after initial locking.

4. After locking the image, you can proceed by developing more links. Note how the roads are more closely aligned with the image after the initial image locking. You can improve the alignment by adding more links. Add links until you reach 10 and then select the Register button to build a new affine transformation model.

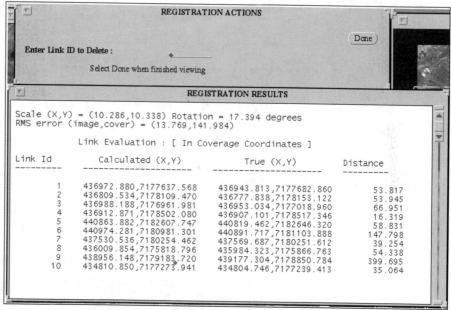

Affine transformations based on 10 links.

Typically, you will see high quality links as well as marginal ones. The affine transformation has an error of 13.769 pixels or 141.984m based on the 10 links. As seen in the Registration Results window, link 9 clearly has the largest error and therefore should be deleted.

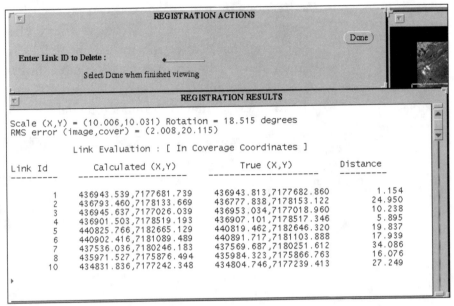

Affine transformations after deleting link 9.

Note that by deleting link 9, the affine transformation RMS error was reduced from 141.984 to 20.115m.

5. You can continue to add and delete links until you are satisfied that the model error is acceptable. When you are satisfied, select the Save Transformation button from the Link Actions menu. Try to develop a model with at least 10 links and an RMS error of less than 1,10 (1 pixel or 10 cover units).

RMS (root mean squared) error is an expression of the model distance error computed from the links' true and calculated positions. An RMS error of 0.0 would mean that the affine transformation perfectly matches the positions of all links that you selected.

➤ **NOTE:** *Frequent clicking on the Save Transformation button in the Link Actions window is recommended. Unfortunately, this action saves affine transformation models only, and not the links used to develop the models.*

✗ **WARNING:** *Clicking on the Reset Transformation button will delete all links!*

Transforming Scanned Maps in ARC/INFO

Scanned maps are in scanner pixel coordinates; therefore, you must use the process described earlier to rectify a scanned map from scanner pixel coordinates to GIS real world coordinates. The process is relatively simple. The first step is to determine the map coordinates of at least four tics on the scanned map. For example, you can use the longitude/latitude tic locations on USGS quadrangles.

In the following exercise, the scanned topographic map, *scanquad.bil* (displayed below), is transformed from scanner coordinates to UTM coordinates. The next illustration shows that TIC 1 has a longitude/latitude of -148° 30', 64° 45'; TIC 2, -148° 20', 64° 45'; TIC 3, -148° 20', 64° 40'; and TIC 4, -148° 30', 64° 40'.

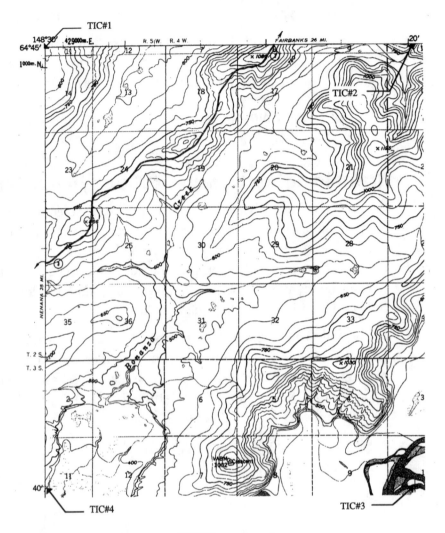

Tic locations on a scanned USGS quadrangle.

1. When you have located at least four tics, you can generate a coverage containing the tics in geographic coordinates. In the following, a coverage is created and tics entered. You must enter the longitude/latitude of each tic as decimal degrees when using GENERATE. For example, TIC 1 would be entered as -148 (30/60) = -148.5000000 64 (45/60) = 64.750000.

```
/***Generate a new coverage called longlat-tics.

Arc: generate longlat-tics

/***Enter the long/lat of the tic locations as seen below.

Generate: ID, X, Y: tics

Generate: ID, X, Y: 1, -148.5000000, 64.7500000

Generate: ID, X, Y: 2, -148.3333333, 64.7500000

Generate: ID, X, Y: 3, -148.3333333, 64.6666667

Generate: ID, X, Y: 4, -148. 5000000, 64.6666667

Generate: ID, X, Y: end /***End of tics.

Generate: quit /***Quit the Generate program.
```

2. Typically your GIS will be in a planar coordinate system and not geographic coordinates. The next step is to project the tics from longitude/latitude to UTM as seen below.

```
/***Project coverage from long/lat to UTM projection.

Arc: project cover longlat.tics utm-tics

Project: input /***Define the input projection.

/***Input projection longitude/latitude.

Project: projection geographic

/***Decimal degree units.

Project: units dd

/***Map datum is NAD27 (Clarke 1886 spheroid).

Project: datum nad27
```

```
Project: parameters /***Done with input projection parameters.

Project: output /***Define the output projection.

/***Universal Transverse Mercator projection.

Project: projection utm

Project: units meters /***Output units in meters.

Project: datum nad27 /***Map datum is NAD27 (Clarke 1886 spheroid).

Project: zone 6 /***Location of topo is in UTM zone 6.

Project: parameters /***Done with output projection parameters.

Project: end /***End of project program.
```

3. Finally, list the tic table to view the UTM coordinates of the four tics.

```
Arc: list utm-tics.tic
```

RecordID	TIC	XTIC	YTIC
1	1	428607	7181244
2	2	436539	7181067
3	3	436343	7171781
4	4	428387	7171959

4. Now use the REGISTER command to develop an affine transformation model and convert the scanned quad from scanner pixel coordinates to UTM coordinates.

```
Arc: register scanquad.bil
```

5. Use the mouse Adjust button to reduce the size of the frame in the Image window. Use the mouse Select button to move the small frame to the upper left corner of the scanned map in the Image window. Next, press the mouse Menu key to display this area in the Overlay window. The simplest way to develop the links is to point to a tic location on the scanned map and then type in its UTM coordinate, as shown in the next illustration.

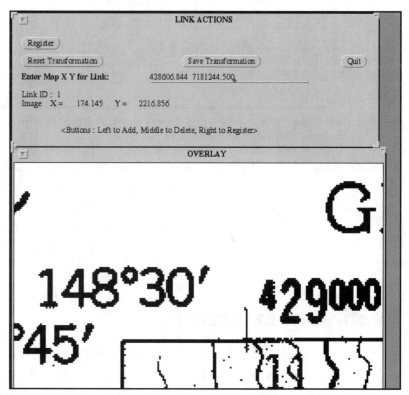

Selecting a tic location and inputting associated UTM coordinates for link 1.

6. Repeat the process of pointing to the tics in *scanquad.bil* and typing in respective UTM coordinates. Then select the Register button to develop an affine transformation.

```
┌─────────────────────────────────────────────────────────────────┐
│ ▽                        REGISTRATION RESULTS                     │
├─────────────────────────────────────────────────────────────────┤
│ Scale (X,Y) = (5.382,5.386) Rotation = 1.526 degrees          ▲  │
│ RMS error (image,cover) = (0.249,1.338)                       ▼  │
│                                                                   │
│           Link Evaluation : [ In Coverage Coordinates ]           │
│ Link Id        Calculated (X,Y)          True (X,Y)       Distance│
│ ---------    -------------------     -------------------   --------│
│         1    428605.756,7181245.281   428606.844,7181244.500  1.340│
│         2    436539.777,7181066.219   436538.688,7181067.000  1.340│
│         3    436342.195,7171781.779   436343.281,7171781.000  1.337│
│         4    428388.086,7171958.221   428387.000,7171959.000  1.336│
│ ◆                                                                 │
│                                                                   │
└─────────────────────────────────────────────────────────────────┘
```

Registration results based on the four map tics. The RMS error represents the average error in the transformation model.

7. If you are satisfied with the transformation RMS error (in the printed example, .249 pixel or 1.338m), click on the Save Transformation button to save the affine transformation models into an image world file. Once you have saved your work, click on the Quit button to exit the REGISTER program.

Rectifying Images

Once you have developed the affine transformation for unwarping, you can use it to rectify an image into a GIS map projection. The process of rectification is analogous to using transformation models to adjust image scale, translation, rotation, and skewness, and then placing the unwarped image over an empty map grid. However, the unwarped image cells will not perfectly match up with the map grid cells.

Resampling Options

Resampling options for filling the empty map grid include nearest neighbor, bilinear interpolation, and cubic convolution. Nearest neighbor is the fastest method, and cubic convolution is the slowest. Because every single map cell in the empty map grid must be filled, image resampling is relatively computer-intensive. Image smoothness increases from nearest neighbor to bilinear to cubic convolution. The three options are briefly described below.

- *Nearest neighbor resampling* fills each map grid cell with the nearest unwarped image pixel values.

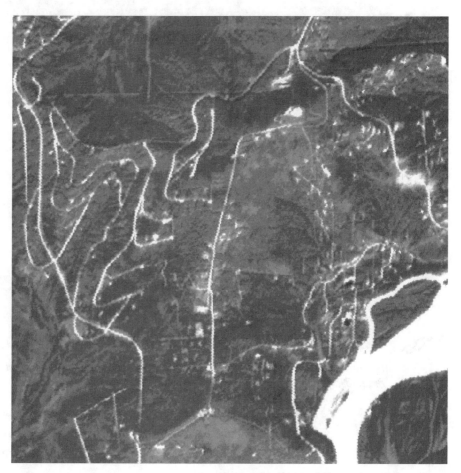

Becker Ridge/Rosy Creek area of Alaska rectified using nearest neighbor resampling.

- *Bilinear interpolation resampling* fills each map grid cell with the weighted average of the four nearest unwarped image pixel values.

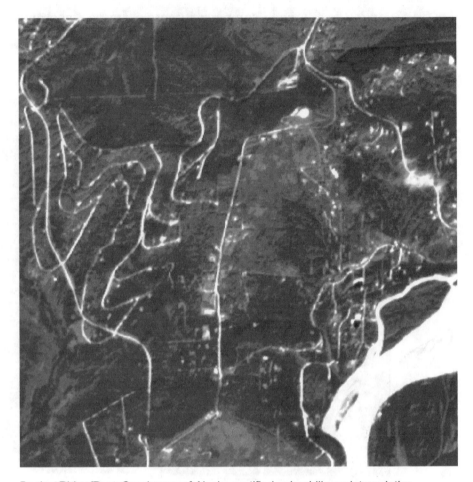

Becker Ridge/Rosy Creek area of Alaska rectified using bilinear interpolation resampling.

- *Cubic convolution resampling* fills each map grid cell with the weighted average of the 16 nearest unwarped image pixel values. The closer a pixel is to the map cell to be filled, the greater its influence in the weighted average value.

Becker Ridge/Rosy Creek area of Alaska rectified using cubic convolution resampling.

Image Rectification Using ARC/INFO

Now that affine transformations for the *spot-pan.bil* and *scanquad.bil* images have been developed, the images can be rectified via ARC/INFO's RECTIFY command. In the following exercise, the *scanquad.bil* image is rectified to the UTM map projection. The *spot-pan.bil* image clipped to the same area is then rectified. The BOX option is used to clip out the rectified satellite image to the same extent as the *scanquad-UTM.bil*.

> ✏ **NOTE:** *Rectifying images takes considerable computer processing time. See Chapter 13 for submitting rectification batch jobs to be automatically executed during off-peak hours.*

```
/***Rectify scanned map image based on affine. Transformation

/***model coefficients are stored in scanquad.blw worldfile. Use

/***nearest neighbor resampling because scan has 1/0 binary pixel

/***values.

Arc: rectify scanquad.bil scan-utm.bil nearest

/***Rectify SPOT satellite image based on affine transformation

/***model coefficients stored in spot-pan.blw world file. Use

/***bilinear resampling because pixel values range 16-100 and

/***weighted averaging is possible. Clip the rectified image to

/***a box.
Arc: rectify spot-pan.bil spot-utm.bil bilinear # BOX
428387,7171781 436539,7181244

/***Examine the rectified images using ARCPLOT.

Arc: arcplot

Arcplot: display 9999 4 /***Full-screen, X-Windows canvas.

/***Scale canvas to fit rectified scan.

Arcplot: mapexent image scan-utm.bil

/***Display the rectified scan image.

Arcplot: image scan-utm.bil
```

```
/***Display the rectified, clipped SPOT satellite image.

Arcplot: image spot-utm.bil

Arcplot: quit /***Quit ARCPLOT module.
```

Rectifying Grids Using ArcView

You can use Avenue scripts to unwarp grids in ArcView GIS if you have the Spatial Analyst extension. In the next exercise, the scan image, *npole.tif*, is converted to a grid. The grid is then rectified to the UTM projection. The image was scanned from a USGS 1:25,000 scale quadrangle. The scan contains noise or speckle which will removed in Chapter 11. Note that the UTM grid was printed on the map; you can use this grid for tic locations. For example, the intersection of the 483 and 7183 lines would have a UTM coordinate of 483000,7183000.

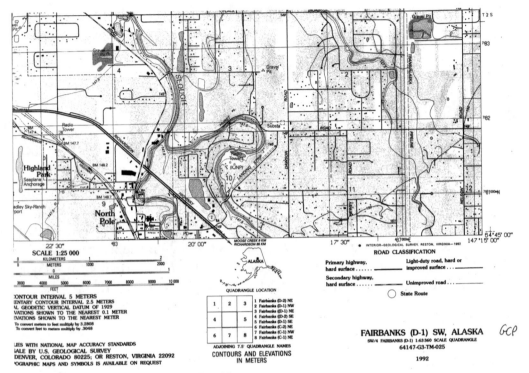

Tic locations on scanned map of npole grid.

⚓ **NOTE:** *You must have the ArcView GIS Spatial Analyst extension loaded to complete the following exercise.*

1. Start ArcView and create a new view. From the View menu, select Add Theme. Add *npole.tif* by double-clicking on it. In order to select the image, the Data Source Type must be set to Image in the Add Theme menu. Once you have added *npole.tif* as an active theme, select the Convert To Grid option under the Theme menu. Enter *npole* as the new grid name and click on the OK button. You will then be asked whether you wish to add the grid as a theme to the view. Click on Yes.

2. Use the Legend Editor to make 0 transparent, and 1 black for the *npole* grid. The next step is to determine the pixel coordinates of four UTM tic locations. Using the magnifying glass icon, zoom in on the intersection of the 483000, 7183000 UTM grid lines. With the Identity tool icon, click on this location. A pop-up table will then display the pixel coordinates for this location at around 657,2300. Now use the Identity tool to find the pixel coordinates on the scan for the UTM coordinates appearing in the next table. For each tic location, estimate the pixel coordinates and the UTM coordinates.

UTM coordinates		Scan pixel coordinates	
483000	7183000	657	2300
487000	71830000	2542	2313
487000	7181000	2547	1369
483000	7181000	662	1356

3. You can use the above information to build four links and then rectify the grid by executing the Avenue script called *rectifyscript.txt*. Build the links, and then select the Untitled project window in the Window menu. Double-click on the Scripts icon to open the Scripts window. In the Script menu, select Load Text File, and double-click on *rectifyscript.txt*.

4. When the text file is loaded, compile it by clicking on the Compile button (check icon). Verify that the *npole* grid is active in the view. To run the script, click on the Run button (to the right of the Compile button) in the Script window.

5. After the script is run, a temporary grid called *npole-UTM* will be added to the view. To save the file as a permanent grid, select the Convert To Grid option under the Theme menu. You will then be prompted for a name for the grid. Input *npole-UTM* to save the rectified grid.

Part IV
Classifying Images

Unsupervised Classification

Once an image has been rectified, it can be classified into a grid of land cover classes. There are two classification strategies: unsupervised and supervised. In unsupervised classification, you group pixels of similar spectral values into *spectral classes*, also known as *spectral clustering*. The next step is to determine what each spectral class represents in terms of land cover. In supervised classification, you outline land cover areas on the image as representative samples. The GIS software then uses the representative samples to predict the land cover for every pixel in the image.

During spectral clustering in unsupervised classification, parameters such as the desired number of spectral classes are specified. The Arc-View GIS Spatial Analyst extension is restricted to single band clustering, while ARC/INFO GRID can execute multi-band clustering on a stack of grids.

The second step in unsupervised classification is to assign a cover type class to each spectral class. This process typically involves tools such as hierarchical cluster analysis to produce an overlay on a color infrared image to determine spatial similarity of spectral classes. Files from the companion CD used in exercises and as references are listed below.

islands.bil	Scanned color, rectified infrared aerial photograph.
islandsc1	Grid corresponding to the channel 1 or band 1 of islands.bil.
islandsc2	Grid corresponding to the channel 2 or band 2 of islands.bil.
islandsc3	Grid corresponding to the channel 3 or band 3 of islands.bil.
islands.blw	World file for islands.bil image.
islands.hdr	Header file describing islands.bil image.
islands.stx	Statistics file for islands.bil image.
yellow.rmt	Remap table for assigning yellow color with GRIDQUERY command.
reclass.rmt	Remap table for reclassifying spectral classes to land cover classes.
cover-types.rmt	Remap table for assigning grid shades to land cover classes.
unsupervised.doc	ARC/INFO documentation file.

In the following sections, unsupervised classification is used to produce an output grid or classified image. The areas to be classified to cover types are part of the Bonanza Creek Long-term Ecological Research Site located in interior Alaska. (For more information on the Site, access *http://www.lter.alaska.edu.*)

Unsupervised Classification Using ArcView

In this section, a scanned, rectified, color infrared aerial photograph called *islands.bil* is classified according to the cover types listed in the following table. The image has been rectified to 5m pixels in the UTM coordinate system.

Cover class	Cover type
1	Water
2	Exposed sediments
3	Muskeg/shrubs
4	Coniferous forest

Spectral Clustering

Take the following steps to classify pixels in the *islands.bil* image.

1. Start ArcView GIS and verify that the Spatial Analyst extension is loaded. Create a new view. Add the *islands.bil* image as a theme by selecting the Add Theme option from the View pull-down menu. In the Add Theme dialog, Data Source Type must be set to Image in order to successfully select *islands.bil*.

2. Add the *islandsc3* grid theme to the view. This grid represents the reflected blue light sensed by the scanner from the photograph. Because the photograph is color infrared, the blue emulsion layer represents the green spectral response. Therefore, in this grid turbid water will have high pixel values, and conifer forest, relatively low pixel values.

3. With the *islandsc3* grid theme active, double-click on the theme to access the Legend Editor. Click on the Classify button, set the Classification Type to Natural Breaks, and specify 10 classes. Click on the OK button to apply the classification rule. Apply Full Spectrum color ramp to assign new colors to the spectral classes.

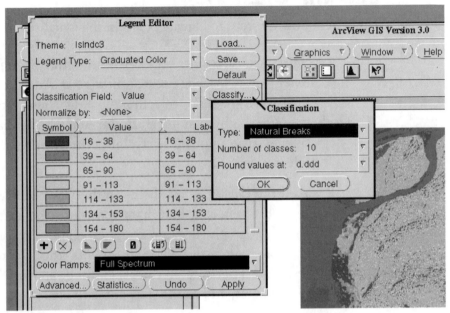

Legend Editor settings for 10 spectral classes under natural breaks classification.

The natural breaks classifier should have produced the spectral classes listed in the next table.

Spectral class	Range of pixel values
1	16-38
2	39-64
3	65-90
4	91-113
5	114-133
6	134-153
7	154-180
8	181-212
9	213-239
10	240-255

Assigning Cover Types to Spectral Classes

The next phase is to assign a cover type to each spectral class. Take the following steps.

1. Use the check box next to the *islandsc3* theme Table of Contents to turn the grid theme, displayed over the color infrared photograph, on and off. From visual interpretation of the color infrared photograph, you can then decide which cover type each spectral class represents. The cover type assignments listed in the next table were derived from visual interpretation.

Spectral class	Dominant cover type
1	Conifers
2	Conifers
3	Muskeg/shrubs
4	Muskeg/shrubs
5	Muskeg/shrubs
6	Muskeg/shrubs
7	Muskeg/shrubs
8	Exposed sediments
9	Water
10	Water

2. Assume that you are satisfied with the cover type assigned to each spectral class and wish to create a new grid using the assignments. Verify that the *islandsc3* grid is active in the Table of Contents. Use the Reclassify function in the Analysis menu to reclassify the active grid to four cover type classes. Select the Classify button to access the Classification dialog box, and change the Number of classes to 4.

Add the following value ranges for each cover type class: cover type 4, 16-64; 3, 65-180; 2, 181-212; and 1, 213-255. Click on the OK button to execute the reclassification.

Reclassify function in Analysis menu.

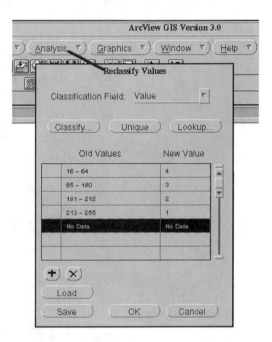

3. Once the reclassification is complete, you will see a new theme titled *reclass of islandc3* in the Table of Contents. Select this theme as the active theme. Change the name of the theme by selecting Properties in the Theme menu. Input *cover types* as the new theme name.

4. Use the Legend Editor in the Theme menu to assign a cover type label and shade color to each class. Finally, use the Legend Editor to turn off the "no data" category in the legend.

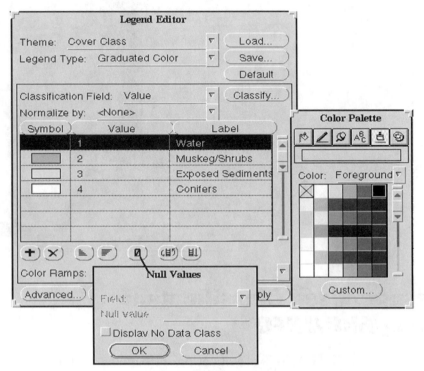

Assigning labels and shades to the cover types grid.

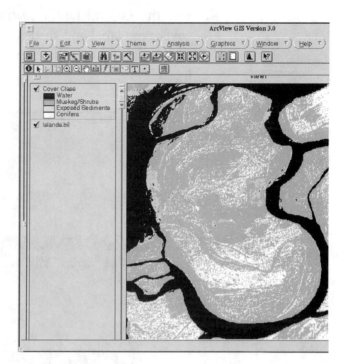

Classified image using grayscale shades assigned to the four cover types.

Unsupervised Classification Using ARC/INFO

ARC/INFO allows classifications using more than one grid or spectral band. In fact, you can use up to 20 grids simultaneously in a classification. When the grids are derived from different spectral regions, the classification is typically called "multispectral."

Single Band versus Multiband Classification

The pixel values from two cover types might not differ significantly using a single grid. For instance, assume that classes 1 and 2 have overlapping pixel values. In this case, accurate classification using a single grid or band would be impossible as shown in the following histograms.

Hypothetical histograms of pixel values from two cover types.

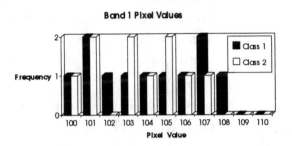

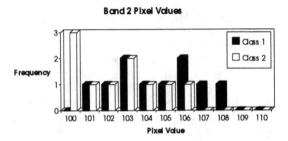

However, if you used both grids simultaneously in the classification, you could correctly separate class 1 versus class 2 pixels as shown in the next illustration.

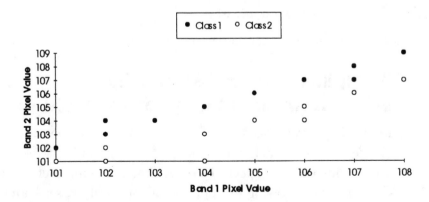

Multiband plot of pixel values from class 1 and 2 cover types.

Spectral Clustering Using ARC/INFO GRID

In this section, the red, green, and blue grids from the *islands* sample scanned aerial photograph are used to experiment with multispectral unsupervised classification. GRID's ISOCLUSTER function is based on the iterative self-organizing data analysis algorithm (ISODATA). (Consult ARCDOC for details on the ISODATA clustering procedure.)

In the following exercise, the ISOCLUSTER function is used to generate 20 spectral classes.

```
/***Make a stack of the three colors scanned from the aerial photo.

makestack islands list islandsc1 islandsc2 islandsc3

liststacks /***List all available stacks.

list islands.stk /***List the grids in the stack called islands.

/***Multispectral unsupervised classification.

/***Generate statistics for 20 spectral classes, 100 iterations,

/***and a minimum of 5 pixels per class.

unsup20 = isocluster(islands, 20,100, 5)

/***Use the grid statistics file to predict the

/***spectral class of each pixel.

20classes = mlclassify(islands, unsup20.gsg)
```

Mapping Spectral Similarity with Hierarchical Cluster Analysis

Multiple spectral classes representing a single cover class are common. Cluster analysis, which builds a map of spectral similarity, is commonly used to identify the spectral classes that represent each cover class. Consider the following example of four spectral classes based on two spectral bands.

Mean Spectral Class Pixel Values

Spectral Class	Band 1	Band 2
1	5	5
2	100	100
3	120	140
4	150	160

The spectral classes can be plotted as seen in the next illustration.

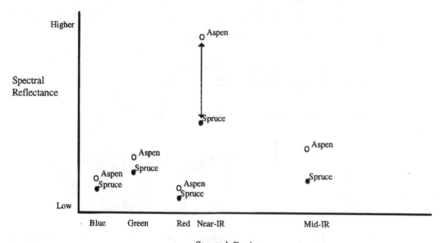

Spectral distance of 134.35 from class 1 to 2.

To calculate the spectral distance between spectral classes, use the following formula.

```
spectral distance = sqrt( (bandX difference)² + (bandY difference)² )
```

The spectral distance between classes 1 and 2 in the previous example is calculated below.

```
sqrt( ( 5 - 100 )² + ( 5 - 100 )² ) = 134.35
```

Using the above formula, the spectral distances between each spectral class are used to compute the following matrix.

Initial Resemblance Matrix

Spectral distances

	1	2	3	4
2	134			
3	177	45		
4	212	78	36	

Based on the previous table, identify the two spectral classes closest to each other. Group the two classes, calculate their mean value, and then fill in a new resemblance matrix.

Second Resemblance Matrix
After Merging Spectral Classes 3 and 4

Spectral distances

	1	2	(3,4)
2	134		
(3,4)	195	61	

Based on the previous table, identify the two spectral classes that are closest to each other. Group the two classes, calculate their mean value, and then fill in a new resemblance matrix.

Final Resemblance Matrix
After Merging Spectral Classes 2, 3, and 4

	1	(2,3,4)
(2,3,4)	174	

Finally, you can create a map of spectral distances based on the resemblance matrices called a *dendrogram*.

Dendrogram or spectral map showing the spectral similarity among spectral classes.

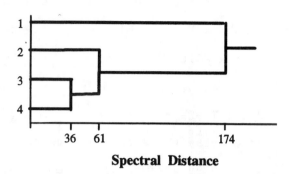

Spectral Distance

Based on the above dendrogram, the two most similar spectral classes that would be candidates for grouping into a single cover class are spectral classes 3 and 4. Spectral class 1 is the most unique. You can generate a dendrogram for the 25 spectral classes using the ARC/INFO GRID module as shown below.

```
/***Mapping spectral similarity

/***Create a dendrogram or map of spectral similarity.

dendrogram unsup20.gsg spectral-map.out

/***System print command to print dendrogram.

&sys print spectral-map.out
```

> ↝ **NOTE:** *The system commands for printing and text editing may be different on your machine than the commands used in this book. Consult with your systems administrator.*

Assigning Cover Types to Spectral Classes

You can now use the dendrogram as a guide to group the spectral classes into cover classes. The GRIDQUERY command can be used with a remap table to overlay spectral classes on the color infrared image. The pound sign (#) in the remap table represents a comment; text following the symbol on a line is ignored in processing.

```
/***Assign cover types to spectral classes.

/***Make a remap table using your text editor.

&sys textedit yellow.rmt

# 7 is yellow when shade set is color.shd.

0 255 : 7 #Remap any pixel with 0-255 values to color 7.

/*** Overlay the spectral classes on the color infrared display.

display 9999 4 /***Full-screen, X-Windows canvas.

/***Scale canvas to fit stack of grids.

mapextent islands

/***Specify shade set to use with gridshades command.

shadeset color.shd

/***Color infrared image.

gridcomposite rgb islandsc1 islandsc2 islandsc3 linear

/***To see through non-query cells.

gridnodata transparent

/***For spectral classes that represent conifer, shade

/***spectral classes 1 through 7 yellow.

gridquery 20classes value yellow.rmt nowrap value in {1->7}

clear /***Clear canvas.

/***Color infrared image.

gridcomposite rgb islandsc1 islandsc2 islandsc3 linear

/***For spectral classes that represent muskeg, shade spectral

/***classes 8 through 11 yellow.

gridquery 20classes value yellow.rmt nowrap value in {8->11}
```

```
clear /***Clear canvas.

/***Color infrared image.

gridcomposite rgb islandsc1 islandsc2 islandsc3 linear

/***For spectral classes that represent shrub/exposed soil, shade

/***spectral classes 12 through 16 yellow.

gridquery 20classes value yellow.rmt nowrap value in {12->16}

clear /***Clear canvas.

/***Color infrared image.

gridcomposite rgb islandsc1 islandsc2 islandsc3 linear

/***For spectral classes that represent water, shade spectral

/***classes 17 through 20 yellow.

gridquery 20classes value yellow.rmt nowrap value in {17->20}

clear /***Clear canvas.

/***Color infrared image.

gridcomposite rgb islandsc1 islandsc2 islandsc3 linear
```

Finally, you can use the RECLASS function to create a grid of cover types based on the spectral classes. First, create a remap table that assigns each spectral class to a cover class.

```
/***Reclassify spectral calss grid to cover type grid.

/***Build remap table for grouping spectral classes.

&sys textedit reclass.rmt

1 7 : 4 #Remap spectral calsses 1->7 to class 4 (water).

8 11 : 3 #Remap spectral classes 8->11 to class 3 (muskeg).

12 16 : 2 #Remap spectral classes 12->16 to class 2 (shrub/soil)

17 20 : 1 #Remap spectral classes 17->20 to class 1 (water).
```

```
/***Reclassify spectral to cover type classes guided by remap table.

cover-types = reclass(20classes,reclass.rmt)

/***Build a remap table for assigning shade colors.

&sys textedit cover-types.rmt

#Assume four cover types and shade set colornames.shd are used.

1 : 43 #Water shaded blue.

2 : 73 #Shrubs shaded green-yellow.

3 : 55 #Muskeg shaded turquoise.

4 : 76 #Conifer shaded forest green.

/***Display classified grid over the color infrared image.

clear /***Clear canvas.

/***Color infrared image.

gridcomposite rgb islandsc1 islandsc2 islandsc3 linear

/***Shade set the cover-types.rmt is designed to use.

shadeset colornames.shd

/***Display classified image.

gridshades cover-types value cover-types.rmt nowrap

quit /***Quit GRID and return to ARC prompt.
```

Supervised Classification

Unsupervised classification assumes that spectral classes are pure representations of land cover, but this is not always true. For instance, the unsupervised classification created in ArcView in Chapter 7 contained some spectral classes which were a mixture of muskeg and shrub. In the ARC/INFO unsupervised classification in the previous chapter, some of the spectral classes were a mixture of exposed sediments and shrub.

In theory, supervised classification can prevent the problem of mixed classes because the analyst selects areas representative of each cover type from an image. These *training areas* are typically digitized polygons of a cover type known to the analyst. The next step is to generate spectral statistics from the pixels inside each training area. This is analogous to showing a GIS what each cover type "looks like" spectrally. The final step is to use the training area statistics to classify the cover type of each pixel in the image. For each image pixel, the classifier compares the pixel statistics with all training area statistics, and then assigns the pixel to the cover type of the spectrally most similar training area.

The first version of the Spatial Analyst extension for ArcView GIS 3.0 does not support supervised classification. Therefore, the exercises in this chapter will be restricted to ARC/INFO GRID. Files from the companion CD used in exercises and as references in this chapter are listed in the following table.

train-polys	Polygon coverage of training areas representing cover types of the sample area.
tmrect2	Rectified grid of Landsat TM band 2 (green region) spectral value.
tmrect3	Rectified grid of Landsat TM band 3 (red region) spectral values.
tmrect4	Rectified grid of Landsat TM band 4 (nearinfrared region) spectral values.
tmrect5	Rectified grid of Landsat TM band 5 (midinfrared region) spectral values.
tm345	Stack listing of themes created in GRID.
tm432	Stack listing of themes created in GRID.
trn-grid.rmt	Remap table for color assignments to training areas.
superv.rmt	Remap table for reclassifying training classes to land cover classes.
colors.rmt	Remap table for assigning grid shades for land cover type classes.
supervised.doc	ARC/INFO documentation file.

Training Area Selection

Training areas are selected as pixels representing each cover type. Therefore, each training area should be as large as possible, while ensuring that all sample pixels are from the given cover type. Training area polygons are typically outlined on the image. Depending on slope gradient and orientation, plant density, plant size, and so on, a cover type may present significantly different spectral responses. Therefore, you should try to develop as many large training areas as possible for each cover type.

In the following example, a relatively large area covered by 25m pixels is classified. The area corresponds to the scanned map image rectified in Chapter 6. The major cover types in the area appear in the next table.

Cover class	Cover type
1	Clear water
2	Turbid water
3	Conifer forest
4	Broadleaf forest
5	Broadleaf shrubs
6	Grasses/burn
7	Muskeg

Using ARCEDIT to Delineate Training Areas

The first step in supervised classification is to delineate training areas on the image. The first step in supervised classification is typically the delineation of training areas on the image. In the next exercise, however, three training areas for each cover class have already been delineated using ARCEDIT (as documented below). Because these training area polygons are saved in the *train-polys* coverage, you can skip the following ARCEDIT statements if you wish. The *tm432* file is also available on the companion CD.

```
Arcedit: display 9999 4 /**Full-screen X-terminal canvas.

Arcedit: mapextent clipcov /***Scale canvas to fit study area.

/***tm432 is a stack listing created in GRID as follows:

/***Grid: makestack tm432 list tm4rect tm3rect tm2rect.

Arcedit: image tm432 composite 1 2 3 /***Stack of tm4,tm3,tm2.

Arcedit: drawenv all /***Set drawenv for editcoverage.

Arcedit: draw /***Draw using the above display rules.

/***Create new coverage with tics from UTM tics.

Arcedit: create train-polys utm-tics

Arcedit: build /***Build initial polygon attribute table.
```

```
/***Add training area polygons on CIR image.

Arcedit: editfeature polygon

Arcedit: add /***Add three polygons for each cover type.

/***1,2,3=clear water, 4,5,6=turbid water, 7,8,9=spruce forest

/***10,11,12=broadleaf forest

/***13,14,15=broadleaf shrubs, 16,17,18=grasses, 19,20,21=muskeg

Arcedit: build /***Build polygon attribute table.

Arcedit: save /***Save work.

Arcedit: quit /***Quit ARCEDIT module.
```

Developing Training Area Statistics

The next step is to develop mean and variation statistics for the pixels within each training area. First, convert the training area polygons to grid cells that overlap the spectral grid stack. Next, extract the pixel values from the grid stack that corresponds to the training area grid cells. The statistics are then computed from the extracted grid pixels. You can use ARC/INFO as follows to accomplish this.

```
/***Convert training polygons to grid.

Arc: polygrid train-polys trn-grid train-polys-id

/***Converting polygons from train-polys to grid trn-grid

/***Cell Size (square cell): 25

/***Convert the Entire Coverage? (Y/N): yes

/***Number of Rows = 370

/***Number of Columns = 285


/***Look at training areas on color infrared image.
```

Arc: grid /***Start the GRID module.

Grid: display 999 4 /***Full-screen, X-Windows canvas.

Grid: mapextent tmrect2 /***Scale canvas to image extent.

/***Color infrared display.

Grid: gridcomposite rgb tmrect4 tmrect3 tmrect2 linear

/***Make a remap table for assigning colors to training areas.

Grid: &sys textedit trn-grid.rmt

#Remap table to color training areas assuming

#shade set is colornames.shd.

1 3 : 46 #Clear water training areas shaded sky blue.

4 6 : 98 #Turbid water training areas shaded chocolate.

7 9 : 76 #Spruce forest training areas shaded forest green.

10 12 : 74 #Broadleaf forest training areas shaded lime green.

13 15 : 83 #Broadleaf shrubs training areas shaded yellow.

16 18 : 110 #Grass/burn training areas shaded red.

19 21 : 127 #Muskeg training areas shaded purple.

/***Select the shade set for which trn-grid.rmt is designed.

Grid: shadeset colornames

/***Set nodata cells to clear.

Grid: gridnodata transparent

/***Shade the training area gridcells.

Grid: gridshades trn-grid value trn-grid.rmt

```
/***Develop training area statistics.

/***Make stack of three spectral bands.

Grid: makestack tm345 list tmrect3 tmrect4 tmrect5

/***Extract statistics from training areas.

Grid: train.gsg = classsig(tm345,trn-grid)

/***Use your system's print command to print statistics.

Grid: &sys print train.gsg
```

With the availability of representative pixel statistics from each training field area, the next step is to use the maximum likelihood rule to classify each pixel based on the statistics.

Maximum Likelihood Classification

The maximum likelihood rule is common to supervised classification. This statistical classifier is based on the assumption that most pixel values for a cover class occur near the mean pixel value of the class. Assume that you wish to classify two cover types: snow and non-snow. Hypothetical statistics for snow and non-snow pixels appear in the next table.

Non-snow pixels		Snow pixels	
Mean	Standard deviation	Mean	Standard deviation
50	10	100	20

The maximum likelihood classifier is optimal when the pixel distribution from each cover class is bell shaped and peaks at the class mean. The statistical formula for a bell shaped (normal) distribution follows:

```
likelihood(PV) = [1/(sd * sqrt(2 * pi))] exp - [(PV - mean)² / (2 * sd²)]
```

PV is the pixel value of a given cover class. The *exp* term is the inverse of the natural log. Also, note that the term $-[(PV - mean)^2 / (2 * sd^2)]$ carries a negative sign. For example, the likelihood of a pixel with a value of 50 belonging to the non-snow class would be as follows:

```
likelihood(50) = [1/(10* sqrt(2 * 3.14159...))] exp - [(50 - 50)² /
(2 * 10²)]
```

```
= 0.03989 exp - [ (0) ]
```

```
= 0.03989
```

Values for non-snow versus snow cover types, calculated using the maximum likelihood rule, are displayed in the next table.

Pixel value	Non-snow likelihood	Snow likelihood
30	54×10^{-4}	
40	242×10^{-4}	2×10^{-4}
50	399×10^{-4}	9×10^{-4}
60	242×10^{-4}	27×10^{-4}
70	54×10^{-4}	65×10^{-4}
80	004×10^{-4}	121×10^{-4}
90		176×10^{-4}
100		199×10^{-4}
110		176×10^{-4}

The previous likelihood values are plotted in the next illustration.

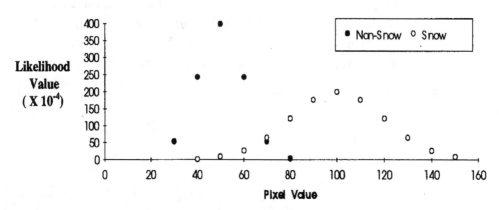

Likelihood curves for snow versus non-snow. The curves intersect at a pixel value of 69.35.

Note that a pixel with a value less than 69.35 has a higher likelihood of belonging to the non-snow than to the snow cover class. Thus, for each pixel in the grid, the GIS can calculate cover class likelihoods and then assign the cover class with the highest likelihood to the pixel.

The maximum likelihood rule is typically applied to more than one grid. For example, the likelihood contours in the next illustration were developed using two spectral bands.

Likelihood contours based on two spectral bands or axes.

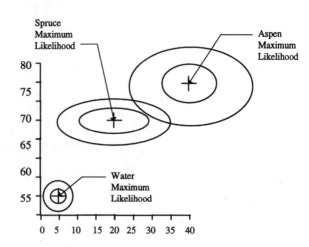

Using the maximum likelihood rule, the previous two-band grid is classified as follows: the class value of 1 represents water, 2 represents spruce, and 3 represents aspen cover types. The following tables display the image's classification according to maximum likelihood.

Original 2-Band Image			
6	5	25	29
44	43	67	60
6	4	25	30
43	43	65	61
10	14	40	43
65	68	77	78
12	15	39	44
66	69	76	78

Classified Image			
1	1	2	2
1	1	2	2
2	2	3	3
2	2	3	3

Classification of two-band image based on maximum likelihood.

Using ARC/INFO GRID

In the preceding section, training area polygons were used to extract representative spectral statistics from the grid cells within each polygon. In the remainder of this section, these training area statistics and the maximum likelihood rule are used to predict the class membership of each pixel.

```
/***Use train.gsg statistics with maximum likelihood classifier.

Grid: superv = mlclassify(tm345, train.gsg)

/***List the classified grid value attribute table.

Grid: list superv.vat
```

The three training areas represent each of the cover types. Therefore, the image is classified. The next step is to reclassify, so that a single value represents each cover type. First, use the text editor to create and save a remap table assigning the class values to each cover type value. Next, use the remap table with the RECLASS function to execute the grid reclassification. The pound sign (#) represents a comment; text following the symbol on a line is ignored by the RECLASS function.

```
Grid: &sys textedit superv.rmt

#Remap table to group classes from 3 training areas per cover type.

1 3 : 1 #Group classes 1,2,3 to class 1 (clear water).

4 6 : 2 #Group classes 4,5,6 to class 2 (turbid water).

7 9 : 3 #Group classes 7,8,9 to class 3 (spruce forest).

10 12 : 4 #Group classes 10,11,12 to class 4 (broadleaf forest).

13 15 : 5 #Group classes 13,14,15 to class 5 (shrubs).

16 18 : 6 #Group classes 16,17,18 to class 6 (grasses/burn).

19 21 : 7 #Group classes 19,20,21 to class 7 (muskeg).
```

Save the remap table.

```
/***Group the supervised classes into the seven cover types.

Grid: class = reclass(superv,superv.rmt)

/***Verify that value attribute table contains 7 classes.

Grid: list class.vat

/***Verify that the reclassification is correct.

Grid: makestack temp-stk list superv class

/***Examine the grid cells to determine if reclassification is cor-
rect.

Grid: cellvalue temp-stk *

/***Press 9 key while in canvas to exit cellvalue program.

Grid: kill temp-stk /***Eliminate the temporary stack list.
```

View the supervised classification. First, display the original color infrared image, and then use a text editor to create a remap table for assigning shades to each class. At this point, you can use the remap table to display the classified grid.

```
/***Color infrared display.

Grid: gridcomposite rgb tmrect4 tmrect3 tmrect2 linear

/***Use text editor to create remap table for assigning shades to
each class.

Grid: &sys textedit colors.rmt

#Remap table for assigning shade colors to class values.

#Assumes colornames.shd shade set is specified.

1 : 43 #Clear water shaded blue.

2 : 55 #Turbid water shaded turquoise.

3 : 61 #Spruce forest shaded dark green.

4 : 73 #Broadleaf forest shaded green yellow.

5 : 97 #Shrubs shaded tan.

6 : 110 #Grass/burn/roads shaded red.

7 : 63 #Muskeg shaded dark sea green.
```

Save the remap table.

```
/***Specify shade set.

Grid: shadeset colornames

/***Use gridshades to display classification.

Grid: gridshades class value colors.rmt
```

You can visually assess the accuracy of a classification, but such practice is rather subjective. Chapter 9 focuses on quantitative classification accuracy assessment.

9

Accuracy Assessment

Thorough accuracy assessment is critical before accepting image classification results. The error matrix is typically the foundation of accuracy assessment because it compares predicted with true cover types. For example, the error matrix for the seven-class grid (*class-grid*) created in Chapter 7 could be structured as shown in the next table.

	Reference Data ("Truth")							
	1 Clear water	**2** Turbid water	**3** Spruce forest	**4** Broadleaf forest	**5** Shrubs	**6** Grasses/burn	**7** Muskeg	**Total**
Predicted								
1								
2								
3								
4								
5								
6								
7								
Total								

Files from the companion CD used in exercises and as references in this chapter are listed in the next table.

ref-grid	Reference grid of true cover types at randomly selected locations.
ranpts	Point coverage containing randomly selected locations.
tm432	Stack listing of themes created in GRID.
class-grid	Classified grid from Chapter7.
tmrect.bil	Landsat Thematic Mapper satellite image.
accuracy.doc	ARC/INFO documentation file.

Because generating an error matrix involves comparing two grids, you will need access to either ARC/INFO GRID or the ArcView GIS Spatial Analyst extension in order to perform the exercises in this chapter.

Reference Data

Reference data are collected as "truth" for comparison with predictions in the error matrix. As suggested below, reference data sampling is not a straightforward procedure. Ideally, reference data are derived from many randomly selected samples independent from training areas. The cover type interpreted for each truth sample is usually interpreted as 100 percent accurate.

Methods of determining ground truth include random sampling of individual image pixels, vector-based GIS polygons, and the use of aerial photography as reference data.

Random Sampling of Individual Image Pixels

With random sampling, you select image pixels and then locate the pixels in the field to determine their actual cover type. However, there are several problems with this approach.

The first problem is that because images are always rectified with some RMS error, map coordinates for each image pixel are an estimate with unknown positional error. Some pixels in a rectified image may be in error by 10m, and others by 100m. This is a discrepancy which renders it impossible to pinpoint individual pixels in the field. Another drawback is the time and expense required to search out random points distributed across remote and rugged landscapes. Finally, most randomly located points might fall into a single cover type. For example, if willow shrubs are relatively rare across the landscape, it would be unlikely that many randomly selected points would occur in this class. However, the staff moose biologist, for instance, may be interested in only the accuracy of willow shrub classification.

Vector GIS Polygons

On occasion, an existing GIS vegetation theme contains polygons that extend across some of the classified image. Using such vegetation polygons for accuracy assessment means savings in time and money toward developing a reference data set. However, using vector vegetation polygons as reference data also has disadvantages. For example, if a significant time lag between the vector polygon source and the satellite image exists, assessment of the grass/burn and turbid water classes may be inaccurate based on the old polygon data.

Another disadvantage is that vegetation polygons are typically visually interpreted to a minimum mapping unit. Therefore, if the minimum mapping area for vector vegetation polygons is one hectare, then many classified small grass openings and ponds would be determined incorrectly using the vector polygon data. A third uncertainty is the inherent positional error between the vector polygons and the satellite image. All three factors can contribute to a serious bias in estimating classification accuracy.

Aerial Photography

With this approach, random points are generated on a color infrared satellite image; the color infrared photos are then used to ascertain the cover type at each point. The advantage of aerial photography is that cover types can be taken at the same time as the satellite imagery. Therefore, it is relatively inexpensive to interpret many sample points. However, the critical assumption is that the interpretation of the aerial photography is 100 percent correct. For more specific vegetation types, an accurate interpretation of aerial photography may not be possible. For example, assume that shrub cover types include the specific classes of willow, alder, or rhubus instead of the general class of broadleaf shrub. In this case, it would be extremely difficult to precisely interpret the aerial photograph. An on-the-ground sampling would probably be necessary.

Every strategy for sampling reference data may present significant problems. Random sampling of individual pixels suffers from high cost, inherent positional uncertainty, and inadequate numbers of rare cover type samples. Use of existing GIS vegetation themes is characterized by inherent positional uncertainty, errors in theme classes, and minimum mapping units larger than a single pixel. Point sampling from aerial photography assumes that positional and visual interpretation errors are minimal. Because of the problems described here, a thorough classification accuracy assessment must report in detail the methods used to sample reference data. Reporting the error matrix alone is insufficient.

Using a GIS to Generate an Error Matrix

Stratified Random Sampling with ARC/INFO

Because the classification of cover types is relatively broad, you can use aerial photography for reference data. A minimum of 30 random sample locations per cover type were generated based on four photographs for the classification grid shown below. The results are copied to a point coverage named *ranpts*. Because the point coverage is stored on the companion CD (i.e., *ranpts*), you may wish to skip the following exercise in GRID to generate the coverage of randomly located points.

```
/***Select 30 random locations from each vegetation type.

/***Start by assigning 0-10,000 randomly to each grid cell.

Arc: grid /***Start GRID module.

/***Set analysis window to area covered by class grid.

Grid: setwindow class

/***Set analysis grid cell size to same as class.

Grid: setcell class

/***Generate random numbers 1 to 10,000 for each cell.

Grid: ran-grid = int( rand() * 10000 + 0.5)

/***Create a separate grid from each class.

Grid: clear = test(class, 'value eq 1') /***All clear water pixels.

Grid: turbid = test(class, 'value eq 2') /***All turbid water pixels.

Grid: spruce = test(class, 'value eq 3')/***All spruce pixels.

Grid: broad = test(class, 'value eq 4') /***All broadleaf forest.

Grid: shrubs = test(class, 'value eq 5') /***All shrub pixels.

Grid: grass = test(class, 'value eq 6') /***All grass pixels.

Grid: muskeg = test(class, 'value eq 7') /***All muskeg pixels.

/***Every pixel in each class is assigned a random number.

Grid: temp-clr = ran-grid * clear

Grid: temp-tur = ran-grid * turbid

Grid: temp-spr = ran-grid * spruce
```

Grid: temp-brd = ran-grid * broadleaf

Grid: temp-shr = ran-grid * shrubs

Grid: temp-gra = ran-grid * grass

Grid: temp-msk = ran-grid * muskeg

/***Now select 30 points from each random class grid.

Grid: List temp-clr.vat

Grid: list ran-clr.vat

1 0 109355 /***Non-clear water pixels.

2	1	1
3	3	2
4	4	1
5	5	2
6	6	1
7	7	2
8	8	1
9	9	2
10	13	1
11	18	1
12	19	1
13	20	2
14	22	1
15	23	1
16	25	1
17	26	1
18	27	1

19	28	1
20	29	1
21	31	1
22	33	1
23	34	3
24	36	3 /***32 pixels with value <=36.

Grid: verify off /***Allow overwriting of temp grid.

/***Randomly selected pixels = 1.

Grid: ran-clr =test(temp-clr, 'value gt 0 and value le 36')

Grid: list ran-clr.vat

Record	VALUE	COUNT
1	0	118055
2	1	32

Grid: Temp = select(temp-msk, 'value gt 0 and value le 6') ran-msk = test(temp,'value gt 0')

Grid: List ran-msk.vat

Record	VALUE	COUNT
1	0	118056
2	1	31

/***Combine all randomly located pixels into a single grid.

Grid: rancells = ran-clr + ran-tur + ran-spr + ran-brd + ran-shr + ran-gra + ran-msk

Grid: temp = select (rancells, 'value gt 0') /***Convert zeros to NODATA.

Grid: rancells = temp

/***Convert randomly located grid cells to a point coverage.

Grid: gridpoint rancells ranpts value

The point coverage *ranpts* now contains at least 30 randomly located points in each of the seven classes in the classified grid called *class-grid*. You could use either ARCEDIT or ArcView to assign the true cover type to each point. This cover type assignment has been carried out for you using the methods discussed in the remainder of the chapter.

Assigning Reference Data Values Using ARCEDIT

You can use ARCEDIT to assign the truth class for each point by interpreting large-scale color infrared photographs for each point location on the satellite image. This task has been carried out in the *ranpts* point coverage. Consequently, you may wish to skip the following steps which assign cover type labels to each reference point and then convert the point coverate to a reference grid.

```
/***Use ARCEDIT to assign the ground truth class for each random pixel.

Arc: arcedit /***Start the ARCEDIT module.

Arcedit: display 9999 4 /***X-windows screen, full-screen canvas.

Arcedit: editcoverage ranpts /***Editcoverage set to coverage containing random points for reference data.

/***Color infrared image in background.

Arcedit: image tm432 composite 1 2 3

Arcedit: drawenv all /***Drawing rule.

Arcedit: draw /***Draw on canvas the image and the editcoverage.

Arcedit: editfeature points /***Specify that you wish to edit points.

Arcedit: &terminal 9999 /***Specify terminal type for forms menu.

/***Use forms menu to assign ground truth class to each point.

Arcedit: forms value

Arcedit: quit /***Exit ARCEDIT and return to ARC.
```

Finally, convert the ground truth points to a reference grid and compare these pixels with the predicted classes.

```
/***Convert ground truth points to reference grid.

/***Convert points from ranpts to grid named ref-grid.

Arc: pointgrid ranpts ref-grid

Cell Size (square cell): 25

Convert the Entire Coverage(Y/N)?: y

Enter background value (NODATA | ZERO): nodata

Number of Rows = 369

Number of Columns = 319

Percentage of Gridded Cells = 100%
```

Producing an Error Table with ARC/INFO

You can use the ARC/INFO GRID COMBINE function to compare the classified image named *class-grid* with the *ref-grid* file from the companion CD.

```
Arc: grid /***Start the GRID module.

/***Compare reference grid with classified grid.

Grid: Accuracy = combine(ref-grid,class-grid)

Grid: list accuracy.vat
```

Record	Value	Count	Ref-grid	Class-grid
1	1	24	6	6
2	2	6	7	6
3	3	25	7	7
4	4	20	3	1
5	5	6	4	5
6	6	6	3	7

Record	Value	Count	Ref-grid	Class-grid
7	7	30	4	4
8	8	24	5	5
9	9	11	1	1
10	10	28	3	3
11	11	2	7	3
12	12	1	7	1
13	13	30	2	2
14	14	1	4	3

Assigning Reference Data Values Using ArcView

The following instructions specify how reference data classes can be assigned using ArcView. The first step is to display the reference data sample points over the color infrared satellite image in ArcView. Subsequent steps are listed below. Compare results with the *ref-grid* file.

1. Create a new view and select View | Add Theme. Select *tmrect.bil* as the image theme to add to the view.

2. Add the point theme of randomly located points. Select View | Add Theme and change the Data Source Type to Feature Source. Select the point theme named *ranpts*.

3. Specify the marker symbol for each randomly located point in the Legend Editor by double-clicking on *ranpts* in the Table of Contents.

Selecting a marker for the random points with the Legend Editor Marker Palette.

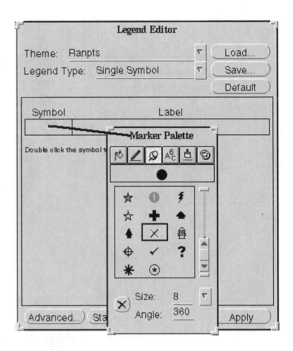

4. The next step would be to assign a cover class to each random point. This step has already been completed. Use the magnifying glass icon to zoom in on an area. Use the Identify tool to examine the cover class interpreted for each random point.

5. To change the cover class assigned to a random point, note the *ranpts#* value of the point and then edit the point attribute table.

6. Select Table | Start Editing. Choose the Change Cell Value icon.

*Changing the cell value
of incorrectly interpreted
cover classes.*

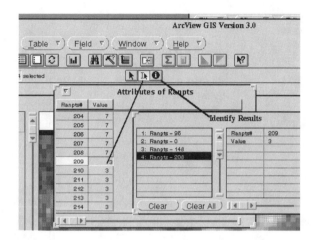

7. When you are satisfied with all cover class assignments to the random points, you could convert the point theme to a grid. Once again, this has already been done for you. To convert a point theme to a grid, make the point theme active and then choose Theme | Convert To Grid.

*Grid conversion
parameters to match
the classified image.*

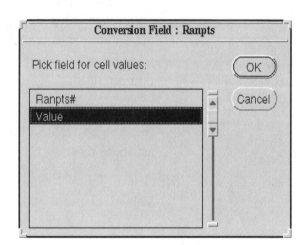

Producing An Error Table Using ArcView Spatial Analyst

To compare the reference grid with the grid cells from the classification, use the Avenue Combine request.

1. Open a new view and add the two grid themes, *class-grid* and *ref-grid*.

2. Select Analysis | Map Calculator. Enter the following Avenue expression in the Map Calculator.

```
( [Class-grid].Combine( { [Ref-grid] } ) )
```

Because the Combine request expects a grid list, *Ref-grid* is enclosed in curly brackets ({}).

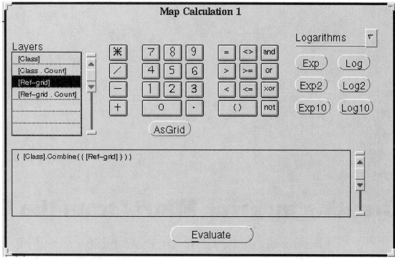

Using the Combine request to compare ref-grid with class-grid.

3. Press the Evaluate button on the Map Calculator to process the Combine expression. The Map Calculation output theme will automatically be added to the view. To view the listing of the output theme table, make the Map Calculation theme active and select Table from the Theme menu. The table should appear as follows.

Count	Truth	Rancells
24	6	6
6	7	6
25	7	7
20	3	1
6	4	5
6	3	7
30	4	4
24	5	5
11	1	1
28	3	3
2	7	3
1	7	7
30	2	2
1	4	3

Creating an Error Matrix from the Error Table

Use the accuracy value attribute table created by using the Combine function in either ArcView or ARC/INFO to fill in the error matrix.

Predicted	Ground Truth							
	1	*2*	*3*	*4*	*5*	*6*	*7*	*Total*
	Clear water	Turbid water	Spruce forest	Broadleaf forest	Shrubs	Grasses/ burn	Muskeg	
1	11	0	20	0	0	0	1	32
2	0	30	0	0	0	0	0	30
3	0	0	28	1	0	0	2	31
4	0	0	0	30	0	0	0	30
5	0	0	0	6	24	0	0	30
6	0	0	0	0	0	24	6	30
7	0	0	6	0	0	0	25	31
Total	11	30	54	37	24	24	34	214

The diagonal cells of the error matrix represent the correct predictions. The overall accuracy of a classification is expressed as the proportion of correct predictions times 100. In the previous error matrix based on the *rancells* reference grid, the total correct prediction is 11 + 30 + 28 + 30 + 24 + 24 + 25 = 172. Thus, the overall classification accuracy is estimated as (172/214) * 100, or 80 percent.

Accuracy Types from the Error Matrix

For each cover type there are two types of prediction error. An *omission error* occurs when a predicted pixel class does not agree with the true class. The accuracy associated with omission errors is referred to as *producer's accuracy*. For example, in the classified image the 11 clear water reference data pixels were correctly classified. The omission error for clear water is zero because no clear water sites from the truth or reference grid were misclassified. Therefore, the associated producer's accuracy for clear water is 100 percent.

A *commission error* occurs when the number of predicted pixels exceeds the true class number. The accuracy associated with commission errors is known as *consumer's accuracy*. In the clear water site example mentioned above, 32 pixels were predicted or committed to clear water. Because only 11 pixels were clear water, there were 32 - 11 = 21 commission errors. The associated consumer's accuracy for clear water was therefore only 11/32 X 100 = 34 percent.

The classification accuracy for clear water is either 100 percent or 34 percent, depending on perspective. Because class accuracy varies widely depending on perspective (user's versus producer's), always report an error matrix when discussing classification accuracy. The reader can then derive producer's and/or user's accuracy estimates for any land cover class in the error matrix.

Class	Omission errors	Producer's accuracy (%)
Clear water	0	11/11 * 100 = 100
Turbid water	0	30/30* 100 = 100
Spruce forest	26	28/54 * 100 = 52
Broadleaf forest	7	30/37 * 100 = 81
Shrubs	0	24/24 * 100 = 100
Grass/burn	0	24/24 * 100 = 100
Muskeg	9	25/34 * 100 = 74

Class	Commission errors	User's accuracy (%)
Clear water	21	11/32 * 100 = 34
Turbid water	0	30/30 * 100 = 100
Spruce forest	3	28/31 * 100 = 90
Broadleaf forest	0	30/30 * 100 = 100
Shrubs	6	24/30 * 100 = 80
Grass/burns	6	24/30 * 100 = 80
Muskeg	6	25/31 * 100 = 81

Part V
Grid Operations and System Integration

Grid Management, Clipping, and Resampling

This chapter is focused on grid operations tools, including grid management, clipping, stitching, and resampling. The following files from the companion CD are used in exercises or as references in this chapter.

ak-dem	Digital elevation grid of most of the state of Alaska.
islandsc1	Grid converted from scanned, rectified aerial photograph.
islandsc2	Grid converted from scanned, rectified aerial photograph.
islandsc3	Grid converted from scanned, rectified aerial photograph.
small-island	Polygon coverage of an island boundary for cutting by polygon exercises.
islands	Stack listing created in GRID.
grid-mgmt.doc	ARC/INFO documentation file.

Management Operations

Operations such as describing, copying, renaming, and deleting grids are relatively simple using ARC/INFO GRID or the ArcView Spatial Analyst extension. The following exercise provides examples of each of these operations using the *islandsc1* grid.

Describing, Copying, Renaming, and Deleting Grids in ARC/INFO

You can use the DESCRIBE, COPY, RENAME, and KILL commands at either the Arc: or Grid: prompt to manage ARC/INFO grids. The following documentation file first acquires all information about the *islandsc1* grid, and then copies it to a grid called *temp*. The *temp* grid is then renamed *trash* and deleted.

```
Arc: listgrids /***You can also use lg.

Workspace: /HOME/IMAGES

Available GRIDs

-------------------

ISLANDSC1

/***Determine grid cell size and projection.

Arc: describe islandsc1

Description of Grid /HOME/IMAGES/ISLANDSC1

Cell Size = 5.000 Data Type: Integer

Number of Rows = 618 Number of Values = 240

Number of Columns = 516 Attribute Data (bytes) = 8

BOUNDARY STATISTICS

Xmin = 438613.218 Minimum Value = 16.000

Xmax = 441193.218 Maximum Value = 255.000

Ymin = 7171280.665 Mean = 168.100

Ymax = 7174370.665 Standard Deviation = 57.054

COORDINATE SYSTEM DESCRIPTION

Projection UTM
```

```
Zone 6

Units METERS Spheroid CLARKE1866

/***Copy the grid to a grid called temp.

Arc: copy islandsc1 temp

Copied islandsc1 to temp

/***Rename temp grid to trash.

Arc: rename temp trash

/***Delete the grid called trash.

Arc: kill trash all
```

Describing, Copying, Renaming, and Deleting Grids in ArcView

The Theme Properties dialog box in ArcView, accessed by selecting Theme | Properties, will display the description of any grid that is an active theme in a view.

Properties dialog box.

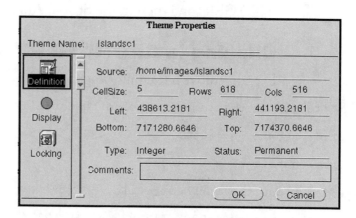

To copy, rename, or delete a grid, open the view window and select File | Grid Manager to open the Grid Manager dialog box.

Grid Manager dialog box.

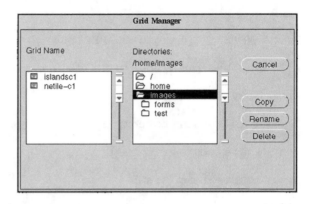

Clipping and Stitching

Several GRID functions can be used to clip a grid into smaller grids. If you wish to clip a grid using a rectangle, you can first specify the rectangle using the SETWINDOW command in GRID. For example, you can split the *islandsc1* grid into four smaller grids by defining a rectangle covering the northeast, northwest, southeast, and southwest portions of the grid. In the following sections, this task is accomplished in ARC/INFO GRID and then in the Spatial Analyst extension.

Clipping in ARC/INFO GRID

Experiment with the following GRID commands.

```
/***Begin by cutting out the northwest tile.

Grid: setwindow 438613,7172825 439903,7174371

Grid: nwtile-c1 = islandsc1

/***Cut out the northeast tile.

Grid: setwindow 439903,7172825 441194,7174371

Grid: netile-c1 = islandsc1

/***Cut out the southeast tile.
```

```
Grid: setwindow 439903,7171280 441194,7172825

Grid: setile-c1 = islandsc1

/***Cut out the southwest tile.

Grid: setwindow 438613,7171280 439903,7172825

Grid: swtile-c1 = islandsc1

/***Evaluate the tiles by displaying them on the canvas.

Grid: display 9999 /***X-Windows, full-screen canvas.

/***Scale canvas to fit extent of original grid.

Grid: mapextent islandsc1

Grid: gridpaint nwtile-c1 /***Display the northwest tile.

Grid: gridpaint netile-c1 /***Display the northeast tile.

Grid: gridpaint swtile-c1 /***Display the southwest tile.
```

Grid: gridpaint setile-c1 value linear # gray /***Display the southeast tile as linear grayscale contrast stretch.

Clipping with ArcView Spatial Analyst

Grid clipping in ArcView GIS is a three-step process. The window extent for clipping is specified, a temporary grid is created from the window area, and the temporary grid is saved to disk.

1. Start ArcView and add a new theme, *islandsc1*, to the view. Verify that the Feature Data Source is set to Grid when you use the Add Theme pop-up menu.

2. When you have successfully added the view *islandsc1* as a grid theme, you can set the extent of the analysis. Verify that *islandsc1* is the active theme. Select Analysis | Analy-

sis Properties. Input the UTM coordinates for the north-west tile as follows: left = 438613, right = 439903, top = 7174371, bottom = 7172825.

Analysis Properties dialog box.

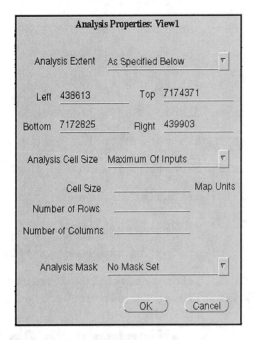

3. At this point, Spatial Analyst knows the extent of the area you want to analyze. Select Analysis | Map Calculator and double-click on the *[Islandsc1]* layer. After you select the Evaluate button on the Map Calculator, a temporary grid called *Map Calculation 1* will be created and added as a theme to the view.

Map Calculator dialog box.

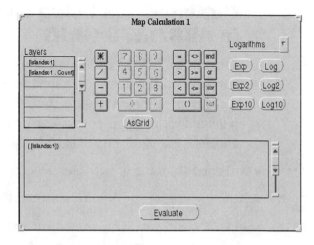

4. The *Map Calculation* theme is a temporary grid. Save this grid to disk so that you can use it later. With the *Map Calculation* theme active, select Theme | Save Data Set. Input *nwtile-c1* as the name of the northwest tile grid to be saved.

Save Data Set dialog box.

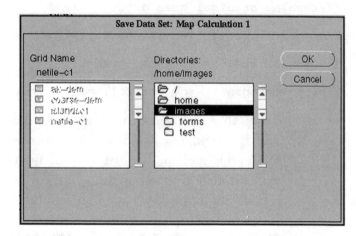

Cutting Grids Using a Polygon Coverage in ARC/INFO GRID

The SELECTPOLYGON function in ARC/INFO GRID is used to cut out areas on the image corresponding to a polygon coverage or a polygon drawn interactively on the image. In the next example, use the *small-islands* polygon coverage to cut out the small islands from the *islandsc1*, *islandsc2*, and *islandsc3* grids.

```
/***Use selectpolygon to cut out small islands from grid.

Grid: smallc1 = selectpolygon(islandsc1,small-island,inside)

Grid: smallc2 = selectpolygon(islandsc2,small-island,inside)

Grid: smallc3 = selectpolygon(islandsc3,small-island,inside)

Grid: mapextent islandsc1

Grid: clear

/***Display original scan grids.

Grid: gridcomposite rgb islandsc1,islandsc2,islandsc3 linear

/***Display the cut out scan grids.

Grid: gridcomposite rgb smallc1,smallc2,smallc3 linear
```

To stitch grids together, use the MERGE command. The first step is to specify the analysis window as the maximum of all input grids. If the input grids overlap, then the order of input grids determines which grid will dominate the overlap zone. In the following example, the four tiles created earlier in this chapter are stitched together.

Stitching Grids in ARC/INFO GRID

```
/***Merge grids.

Grid: setwindow maxof

Grid: whole-c1 = merge(nwtile-c1,netile-c1,swtile-c1,setile-c1)
```

```
/***Display the stitched tiles grid as linear grayscale contrast.

Grid: gridpaint whole-c1 value linear # gray
```

Cutting and Stitching Grids Using a Polygon in ArcView

At present, ArcView Spatial Analyst does not contain menu-based utilities for stitching or selecting polygon operations. Carrying out these operations is possible, but only through an Avenue script. The following table lists the Avenue requests and syntax for these operations.

Operation	Avenue request	Example
Stitching grids	Merge	mergedGrid = aGrid.Merge(GridList)
Cutting grids with a polygon	ExtractByPolygon	clippedGrid = aGrid.ExtractByPolygon(aPoly, aPrj, selectOutside)

Resampling

Resampling is analogous to placing a coarse grid on top of an original fine grid and then filling the coarse grid with values sampled from the fine grid. Resampling methods are listed below.

- Nearest neighbor—For each coarse grid cell, the values from the fine grid are evaluated. The fine grid value that is closest to the center of each coarse grid cell is selected.
- Bilinear interpolation—Weighted mean of the four closest pixel values.
- Cubic convolution—Weighted mean of the 16 closest pixel values.

The bilinear interpolation and cubic convolution methods produce a grid with a smoother appearance compared to nearest neighbor resampling. Cubic convolution produces the smoothest grid, but is also the slowest of the three methods because it computes the weighted mean of 16 pixel values for every pixel in the output grid.

Nearest neighbor resampling, however, has the following advantages over the other resampling methods: (1) speed; (2) it is the only rational choice when resampling grids that contain classes or categories instead of quantities; and (3) it retains the original cell values, while a resampling method based on weighted means could produce values not present in the original grids.

Why would you want to resample a grid? First, you can significantly reduce disk storage space. For instance, assume that a digital orthophoto of 1m pixels requires 100 Mb of storage space. If 2m pixels are acceptable for the application in process, you could resample the grid to 2m pixels, resulting in a file size of approximately 25 Mb. The advantages of resampling in this case are faster display and analysis of the new 25 Mb orthophoto.

Grid resampling will also save time in 3D perspective viewing or analytical hillshading applications. For example, assume that you want to produce a 3D perspective view of the state of Alaska using the *ak-dem* grid. You could save time by resampling the grid to a coarser grid and using it for draft viewing. When you are satisfied with the viewing parameters, you could use the original grid for presentation quality viewing. In the following exercise, the *ak-dem* grid is resampled from 1km cells to 10km cells to create a coarser elevation grid for relatively quick draft 3D perspective or hillshade viewing.

> ↦ **NOTE:** *As discussed in Chapter 6, "Image Rectification," resampling is also required to rectify images.*

Grid Resampling Using ARC/INFO

Use the RESAMPLE command to create a coarser grid from the *ak-dem* file. The syntax for using this function is *output grid = resample(input grid, resample method)* where *resample method* is the keyword *nearest*, *bilinear*, or *cubic*. In the following example, *ak-dem* is resampled to a 10km digital elevation grid. Compare the time it takes to display both the original and resampled grids as 3D perspective views.

```
Grid: display 9999 4 /***Full-screen, X-Windows canvas.

/***Scale canvas to fit Alaska elevation grid.

Grid: mapextent ak-dem

/***Resample 1km grid to 10km grid cells.

Grid: coarse-dem = resample(ak-dem,10000,nearest)

/***Relatively fast surface display.

Grid: clear /***Clear canvas.

Grid: surface lattice coarse-dem

/***Define surface using 10km grid.

Grid: surface lattice coarse-dem

/***Set default viewing parameters.

Grid: surfacedefaults

/***Drape the surface.

Grid: surfacedrape gridpaint coarse-dem value equalarea nowrap gray

/***Relatively slow surface display.

Grid: clear /***Clear canvas.

Grid: surface lattice ak-dem

/***Define surface using 1km grid.

Grid: surface lattice ak-dem

/***Set default viewing parameters.

Grid: surfacedefaults

/***Drape of finer surfaces.

Grid: surfacedrape gridpaint ak-dem value equalarea nowrap gray
```

Grid Resampling Using ArcView

Grid resampling in the ArcView Spatial Analyst is a three-step process. First, you must specify the grid to resample, the output cell size, and the resampling method (nearest neighbor, bilinear, or cubic convolution). Next, create a temporary coarse grid using the .Resample request. Finally, you can save the temporary coarse grid to disk.

1. Start ArcView and add the *ak-dem* grid theme to the view.

2. Select Analysis | Map Calculator. Enter the following expression:

([Ak-dem].Resample (10000,#GRID_RESTYPE_NEAREST))

The expression requests nearest neighbor resampling from *ak-dem* to produce a 10km cell grid. When you select the Evaluate button on the Map Calculator, a temporary grid called *Map Calculation* is created and added as a theme to the view.

Expression in Map Calculator.

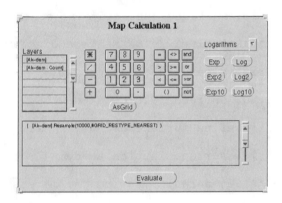

➥ **NOTE:** *The three resampling options are* #GRID_RESTYPE_ NEAREST, #GRID_RESTYPE_BILINEAR, *and* #GRID_RESTYPE_CUBIC.

3. The *Map Calculation* theme is a temporary grid. Save this grid to disk so that you can use it later. With the *Map Calculation* theme active, select Theme | Save Data Set. Input *coarse-dem* as the name of the coarse elevation grid to be saved.

Save Data Set dialog box.

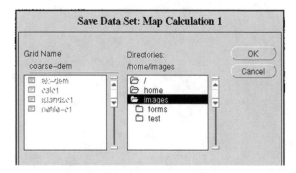

4. Try analytical hillshading from the Analysis menu using *ak-dem* and *coarse-dem*. You should find that the hillshading is much faster with the *coarse-dem* grid. Consequently, you can use *coarse-dem* as a draft grid until you are satisfied with the azimuth and altitude settings. Then use *ak-dem* for the presentation quality hillshade.

Relatively fast analytical hillshading using coarse-dem grid.

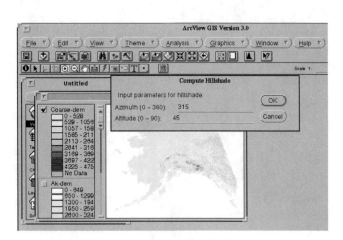

Noise Removal
with ARCSCAN Tools

This chapter is focused on ARCSCAN tools to remove noise and correct misclassified pixels. ARCSCAN was employed in Chapter 3 to perform a heads-up digitizing exercise. Because the scan was clean, raster
editing in ARCSCAN was not necessary. In many scanning applications, however, raster editing is necessary before you attempt to
autotrace vectors from the scan. ARCSCAN is licensed as a separate
ARC/INFO module. See instructions in Chapter 3 for checking on
ARCSCAN license status.

The following files from the companion CD are used in exercises and as
references in this chapter.

npole.tif	Scanned topographic map for the North Pole, Alaska, quad.
npole	Grid contaminated with speckles.
class-grid	Classified land cover grid used in grid editing exercise.
class.rmt	Remap table for assigning colors to land cover exercises.
gridedit.doc	ARC/INFO documentation file.

Removing Speckles

Speckles are small clusters of pixels or noise on a scan that did not appear on the original map. Speckles can be introduced during scanning in at least three ways. First, the quality of the scanned map may be poor, such as fold creases, stains, dirt particles, and so on. Second, a dirty scanner, like a dirty map, will also cause speckles in a scanned data file. Third, the scanning threshold may be set too high, causing the binary scan to include some dithered map colors that appear as speckles. For example, in a binary scan of a colored topographic map, some of the forested areas with green dithering or water areas with blue dithering might appear as speckles on the binary scan.

If you must scan from a map that is not clean and clear, you have at least three options. First, you can retrace the map onto a clear mylar sheet and then scan the mylar, such as the scanned map preparation in Chapter 3. Second, you could scan the map, rectify the scan, and then use it as a background image for on-screen digitizing in ARCEDIT. Third, you can scan the map and use ARCSCAN to clean up the scanned data file for ARCSCAN autotracing. The method depends on the quality of the base map, operation setup, and your preference.

The next exercise uses part of a scanned topographic map from the North Pole, Alaska, at about 15 miles east of Fairbanks. The place is famous for a wood sculpture of a giant mosquito, among other things. The *npole* grid contains speckles as numerous as mosquitoes in the North Pole area during the summer months.

1. Start Arctools.

    ```
    Arc: arctools edit
    ```

2. In the Edit Tools window, select File | Grid | Open. Choose *npole* and click on OK to display the grid for editing on the ARCEDIT canvas. The Grid Editing dialog appears.

Grid Editing dialog box.

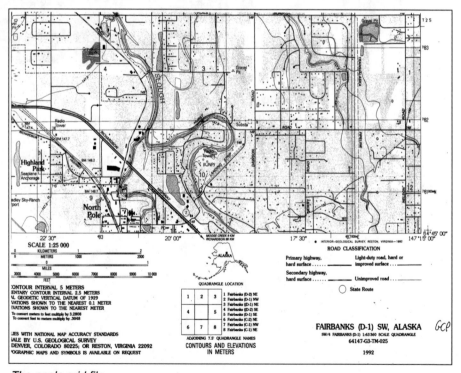

The npole grid file.

3. Use Ctrl+E keys or choose Extent from the canvas Pan/ Zoom menu to clearly distinguish between speckles and valid data, such as buildings.

Examples of speckles versus buildings in npole grid.

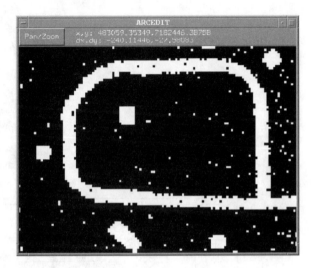

Because distinguishing speckles from valid data is sometimes difficult, using the base map (from which the scan was derived) as a reference is recommended. The USGS quad for the North Pole, Alaska, appears below and as an *npole.tif* image file on the CD for your reference.

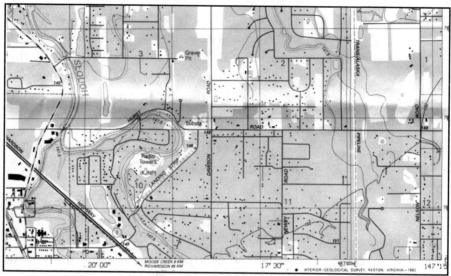

North Pole, Alaska, USGS quad.

4. To remove speckles, you must first inform ARCSCAN of what the speckle "looks like" in terms of maximum pixel width and height. Click the Noise button in the Grid Editing menu to open the Reduce Noise dialog box.

Reduce Noise dialog box.

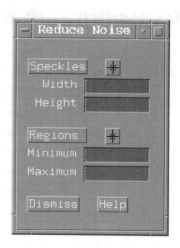

5. Define the maximum size of a speckle as 7 pixels wide and 7 pixels high by inputting these values in the Width and Height fields. Next, inform the module of the extent of the grid from which to eliminate speckles. In this case, you want to eliminate speckles from the entire grid, so click the Selection All button in the Grid Editing menu.

6. Instruct ARCSCAN to eliminate all contiguous pixels that are equal to or less than 7 pixels wide or 7 pixels high by clicking on the Speckles button in the Reduce Noise dialog box.

7. Define speckles as 10 pixels wide and 10 pixels high, and then click the Speckles button to instruct ARCSCAN to eliminate any contiguous pixels that are less than or equal to 10 pixels in width or height.

It is possible that valid data such as small broken lines would also be removed in a Selection All task such as in the previous exercise. An alternative is to apply noise removal to a selected box area rather than the entire map.

Removing Noise Regions

A *speckle* is defined in ARCSCAN as a group of contiguous pixels of identical value that are less than or equal to a user-specified width and height. A *region* is a group of contiguous pixels of identical value within a range of number of pixels in the region. Noise regions are sometimes phantom graphics. Examples include elevation values above contour lines, or coordinate values printed near a lake shoreline that you want to autotrace. Phantom graphics are typically easy to remove as noise regions. For example, try removing the value 149 located in the upper left of the grid.

1. Use Ctrl+E or choose Extent from the canvas Pan/Zoom menu so that the value 149 is clearly displayed on the canvas.

Phantom graphic to be removed with the Regions tool.

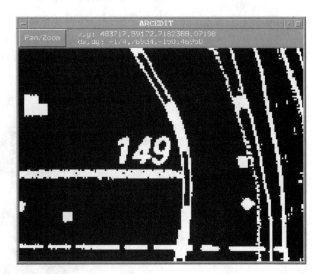

2. To define a noise region for ARCSCAN, first click the graphic button to the right of the Regions button in the Reduce Noise menu.

3. Next, define what a region is in terms of minimum and maximum number of contiguous pixels. Make a box around the number 149 with the mouse. The minimum (90) and maximum (252) numbers of continguous pixels that ARC-SCAN finds inside the box will be displayed below the Regions button. In this example, the number 1 is composed of 90 contiguous pixels, while the number 4 is composed of 252 contiguous pixels.

4. Inform ARCSCAN of the grid extent in which to search for noise regions to be eliminated. Click the Select by Box button in the Grid Editing menu, and define the selected area around the number 149. A green box appears to show you the area extent for editing.

5. Click the Regions button. ARCSCAN will search within the selection box for all groups of pixels sized between 90 and 252 contiguous pixels. The appropriate pixels are then eliminated, and the number 149 disappears from the ARCEDIT window. If you want to try noise removal again, simply click on the Oops button to back up a step.

The npole grid after removal of the number 149.

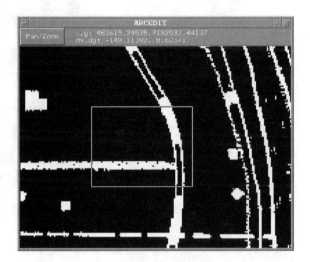

The Regions tool should be used carefully. The Regions tool removes any objects in the selected area that have numbers of cells between the minimum and maximum values listed in the Reduce Noise menu. Therefore, it can remove objects that you actually want to save, such as broken but valid lines. Again, a safe way to use the Regions tool is to work with selected areas rather than the entire map. Remember that you can use the Oops button to reverse any mistakes you make while executing noise removal.

ARCSCAN Sketch Tools

The automated tools to remove speckles and noise regions are useful for working with noise that is isolated from line features you are interested in. To eliminate noise pixels connected to pixels you wish to save, however, you need to use the sketch tools.

Click the Sketch button to open the Sketch menu. The first four options are tools for drawing objects: circle, box, polygon, and line. The next four are tools for filling objects: circle, box, polygon, and pencil. The results of drawing or filling depend on the fill value. There are two possible fill values for binary scans: 1 for objects, and 0 for the background.

Sketch dialog box.

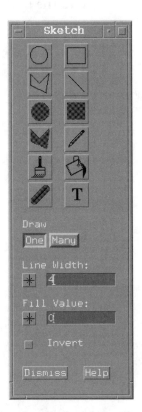

The paintbrush tool works like a paintbrush, brushing the path it passes with the brush fill value and width. The paint bucket tool at the right of the paintbrush is used for filling contiguous same-value pixels with the current fill value. The pencil tool is used to fill individual grid cells one at a time. Finally, the tool at the far left is used to rasterize arcs from a coverage to be added to the grid, and the text tool on the right is used to add text. The line width, measured in number of cells, controls the width of the line and paintbrush tools.

1. Zoom into the area around a gravel pit in the upper middle part of *npole*. Change the Fill Value in the sketch menu to 0. Click the Fill Box tool, and drag the mouse to make a box around the gravel pit symbol. The symbol disappears into the background when the mouse button is released.

*Gravel pit symbol
(a phantom graphic)
to be removed.*

Gravel pit scan after
removal of the gravel
pit symbol.

2. Zoom into the area around the word "SLOUGH." Take a close look at the letters O and U. The letters are connected to lines on both sides.

The word SLOUGH to be
disconnected.

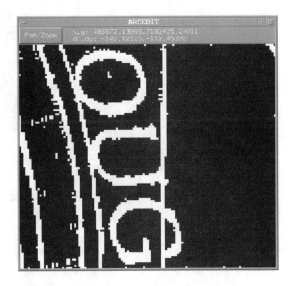

To remove the letters, you must first disconnect the letters from the lines. The Draw Line tool is recommended for this task. In the Sketch menu, change the Line Width to 2 and the Fill Value to 0. Click on the Draw Line button. Left click the mouse at the starting point of a line to disconnect the letter, continue left clicking to make the separation line, and right click the mouse to finish. You should see a clear separation between the letter and the surrounding line. If you make a mistake, click the Oops button to restore the grid to the state prior to the last edit.

Disconnection of SLOUGH from the surrounding lines.

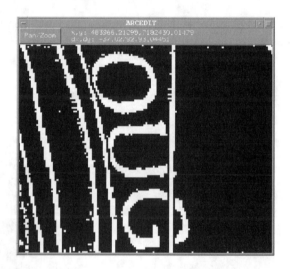

3. After the letters have been disconnected, you can use the paint bucket to flood the letters with a fill value of zero. With the fill value set to zero, select the paint bucket and then click on any point within each letter until the word is removed.

Removal of letters O and U.

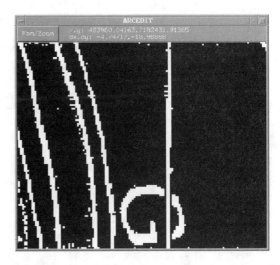

4. Zoom into the area around a body of water to the east of the word SLOUGH. There are speckles inside the body of water caused by scanning the blue dithering on the base map.

Speckles in body of water.

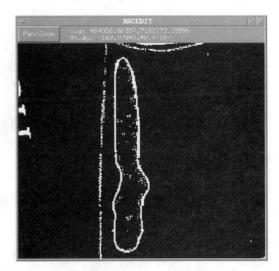

5. Try using the paintbrush to remove the speckles. Change the Line Width to 4 pixels and the Fill Value to 0. Click the paintbrush and move to the canvas. Click the mouse at a point along the inside shore line, and drag the mouse along the entire shore line. The paintbrush remains "on" with the current fill value as long as you continue to hold the mouse button down. Because the paintbrush follows a freehand motion, it is easy to paint over the shore line. The line width determines the width of the stroke of the paintbrush; wider strokes speed up the editing but tend to create more mistakes. For example, try using the paintbrush with the line width set to 20 pixels.

Speckles removed from the body of water.

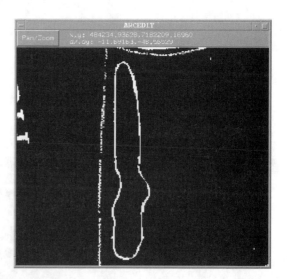

Interactive Image Classification

Results of the accuracy assessment of a land cover classification in Chapter 9 indicated that many of the spruce pixels were misclassified as water. You can use the ARCSCAN sketch tools to correct these misclassified pixels. All ARCSCAN utilities are available through ARC-

TOOLS as well as through ARCEDIT GRIDEDIT. The following documentation file uses these tools in ARCEDIT to change spruce pixels that were incorrectly classified as water (class 1) to the correct class (class 3).

```
/***Make a stack in GRID for color infrared bands.

Arc: grid

Grid: makestack cir list tmrect4 tm4ect3 tmrect2

quit /***Exit GRID module.


/***Display color infrared image.

Arc: arcedit

Arcedit: /***Full-screen, X-Windows canvas.

Arcedit: display 9999 4

/***Scale canvas to fit your study area.

Arcedit: mapextent image tmrect3

/***Display rule for color infrared image.

Arcedit: image cir composite 1 2 3

/***Set drawing environment for editcoverage.

Arcedit: drawenv all

/***Draw using the above display rules.

Arcedit: draw


/***Fix spruce pixels (actual class = 3) that were incorrectly

/***classified as water pixels (predicted class 1)

/***Specify the grid to be edited.
```

```
Arcedit: gridedit edit class-grid

/***Set oops on for backing out of mistakes.

Arcedit: gridedit oopson

/***Edit a remap table for pixel color assignment.

Arcedit: &sys textedit class.rmt &

Arcedit: 1 : 4 #Clear water (1) will be colored blue (4).

Arcedit: 2 : 5 #Turbid water (2) will be colored cyan (7).

Arcedit: 3 : 3 #Spruce (3) will be colored green (3).

Arcedit: 4 : 7 #Broadleaf forest (4) will be colored yellow (7).

Arcedit: 5 : 8 #Broadleaf shrubs (5) will be colored orange (8).

Arcedit: 6 : 14 #Grass/burn (6) will be colored dark gray (14).

Arcedit: 7 : 10 #Muskeg (7) will be colored pale green (10).

/***Specify remap table to use.

Arcedit: gridedit remap class.rmt

/***Draw using display rules

Arcedit: draw

/***Specify fill value for spruce pixels.

Arcedit: gridedit fillvalue 3 /***3 for spruce.

/***Two-pixel paintbrush width.

Arcedit: gridedit linewidth 2

/***Paint incorrectly classified water pixels 9=end.

Arcedit: gridedit brush many

/***Save grid as new grid called corrected.

Arcedit: gridedit save corrected

Arcedit: quit /***Exit ARCEDIT.
```

Filtering and Eliminating Groups of Pixels

This chapter is focused on tools filtering grids and eliminating grouped pixels. These tools are commonly used to enhance the edges on images and smooth grid classifications. The following files from the companion CD are used in exercises and as references in this chapter.

ortho.bil	Digital ortho for edge enhancement exercises.
ortho.blw	World file for digital orthophoto.
ortho.hdr	Header file describing digital orthophoto.
ortho.stx	Statistics file for digital orthophoto.
class-grid	Classified grid of land cover to be filtered or smoothed.
spot-pan.bil	HRV satellite image by SPOT IMAGE. Copyright © CNES/SPOT IMAGE.
filter.doc	ARC/INFO documentation file.

Grid Filtering

A grid filter applies a moving window or kernal to a grid. The output grid is computed from a focal or per neighborhood function applied to the moving window. For example, if you were to specify a three-column by three-row moving window and a focal range function, each pixel in the output grid would contain the maximum range of cell values within a 3x3 window.

ORIGINAL GRID

200	110	210
150	100	170
140	130	160

DUPLICATE BORDER CELLS

200	200	110	210	210
200	200	110	210	210
150	150	100	170	170
140	140	130	160	160
140	140	130	160	160

$\rightarrow$

COMPUTE MAX-MIN

200	210	210
-100	-100	-100
200	210	210
-100	-100	-100
150	170	170
-100	-100	-100

$\rightarrow$

OUTPUT GRID

100	110	110
100	110	110
50	70	70

Focal range function (max-min) applied with a 3x3 moving window.

Edge Enhancement

The objective of edge enhancement is to improve the visual contrast of edges in an image. Typically, edge enhancement is accomplished by applying a filter that returns high values for pixels that are heterogeneous in the image. For example, the edges on the SPOT HRV satellite image, *spot-pan.bil*, could be enhanced by using a range filter.

Original SPOT HRV satellite image. Copyright © CNES/SPOT IMAGE.

Same image as after applying a 3x3 range filter.

In the following exercises using ARC/INFO GRID and the ArcView Spatial Analyst extension, an edge enhancement filter is applied to the *ortho.bil* image that you worked with in Chapter 4.

Edge Enhancement Using GRID

In GRID, all filter functions start with the word "focal." Try applying the FOCALRANGE function using a 3x3 moving window to the digital ortho. The first step is to convert the image to an ARC/INFO GRID format.

```
/***Convert from BIL image to GRID format.

Arc: imagegrid ortho.bil grid-ortho

Arc:grid /***Start the GRID module.

Grid: display 9999 4 /***Full-screen X-Windows canvas.

/***Scale canvas to fit the digital ortho.

Grid: mapextent grid-ortho

/***Grayscale display of orthophoto.

Grid: gridpaint grid-ortho value linear nowrap gray

/***Apply focal range 3x3 filter to enhance edges.

Grid: edges = focalrange(grid-ortho,rectangle,3,3)

/***Grayscale display of edge enhanced orthophoto.

Grid: gridpaint edges value linear nowrap gray
```

Edge Enhancement Using ArcView GIS Spatial Analyst

You can use the Spatial Analyst FocalStats instance request with the Map Calculator to apply a focal range function to a digital orthophoto.

1. Start ArcView and verify that the Spatial Analyst extension is loaded. Create a new view and select View | Add Theme. Select *ortho.bil* as the image theme to add.

2. With *ortho.bil* as the active theme, select Theme | Convert To Grid. Name the output grid *orth-grid* and answer Yes when asked if you want to add this grid theme to the view. Use the Legend Editor to apply a gray monochromatic color ramp to the grid display.

3. With the grid theme *orth-grid* active, select Analysis | Map Calculator. In the Map Calculator dialog box, enter the following expression:

```
( [Grid-ortho].FocalStats(#GRID_STATYPE_RANGE,NbrHood.MakeRectan-
gle(3,3,False),True) )
```

The FocalStats request in the expression requires the following three parameters:

- Type of statistic (min-max range): #GRID_STATYPE_RANGE.

- Size of the moving window: NbrHood.MakeRectangle (3,3,False). A 3x3 moving window is returned, and False means that pixels rather than map units are used in sizing the window.

- If NODATA values are encountered, should they be dominant (yes): True.

4. Click on the Evaluate button to evaluate the expression after you enter it using the Map Calculator. The Map Calculator will add the filtered grid to the view as a temporary grid theme called *Map Calculation 1*. To save the theme as a permanent grid, select Theme | Save Data Set.

Filters For Smoothing Grids

Smoothing filters are often applied to classified images to eliminate scattered single-pixel classes. The majority filter, commonly used for smoothing, replaces the center of a 3x3 moving window with the majority of the contiguous classes surrounding each center pixel. For example, in the following illustration, column #2, row #2 is surrounded by seven contiguous 2 values and therefore receives the majority value of 2. The column #3, row #2 pixel is not surrounded by at least five contiguous pixels of the same value and therefore remains unchanged with a value of 3 in the output grid.

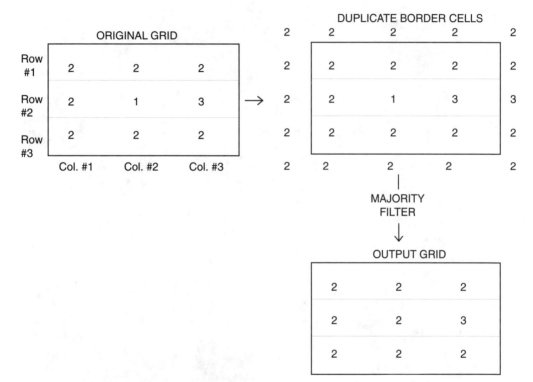

Original grid and output grid after processing by majority filter.

➼ **NOTE:** *Smoothing a classified image with a majority filter may be undesirable for some applications, such as vegetation analysis. Using the majority filter for such analyses would remove information about vegetation patch heterogeneity. For example, wildlife biologists are often interested in the* *dominant vegetation type within a stand, as well as heteroge-neity within the stand.*

The following exercises use ARC/INFO GRID and the ArcView GIS Spatial Analyst extension, to apply the majority filter to the *class-grid* classified image from Chapter 7.

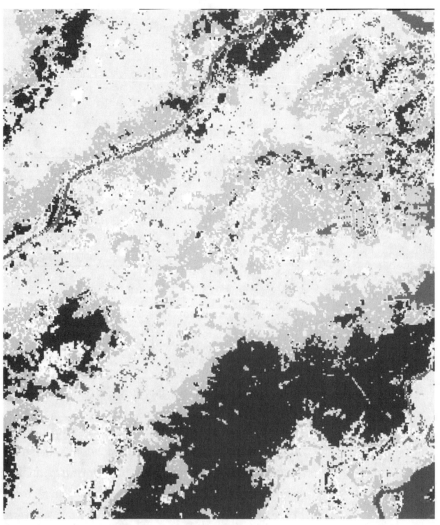

Original classified image.

Above classified image after smoothing.

Majority Filtering Using GRID

Use the following documentation file to smooth out the classified image used in Chapter 7. In the example, the majority filter is applied five times.

```
/***Display original classified grid.

Grid: mapextent class-grid

Grid: gridshades class-grid

/***Smooth the grid with majority filter

/***considering all eight neighboring pixels.

Grid: smooth = majorityfilter(class-grid,eight)

Grid: smooth2 = majorityfilter(smooth,eight)

Grid: smooth3 = majorityfilter(smooth2,eight)

Grid: smooth4 = majorityfilter(smooth3,eight)

/***View smoothed grids.

Grid: gridshades smooth

Grid: gridshades smooth4
```

Majority Filtering Using GIS ArcView Spatial Analyst

In the following exercise, the .MajorityFilter request is used with the Map Calculator to smooth the classified image.

1. Create a new view and add the *class-grid* theme by selecting View | Add Theme.

2. With the grid theme *class-grid* active, select Analysis | Map Calculator. The .MajorityFilter request syntax appears below.

   ```
   aGrid.MajorityFilter(diagNbrs, halfOK)
   ```

   The .MajorityFilter request requires the following parameters:

   - aGrid is the input grid that to be smoothed.

- Set the diagNbrs flag to True if you wish to include diagonal cells in the filter's calculations. Set the flag to False if you want only the upper, lower, left, and right cells considered in the moving window.

- Set the halfOK flag to True if you wish four continguous cells of the same value to be considered a majority. Set the flag to False if wish a majority to be at least five contiguous cells of the same value.

3. Enter the following expression in the Map Calculator dialog:

```
( [Class-grid].MajorityFilter(True, False) )
```

4. Click on the Evaluate button to execute majority filter smoothing on the classified grid.

5. Save the smoothed grid. With the *Map Calculation 1* theme active, select Theme | Save Data Set. Name the grid *smoothed*.

Eliminating Groups of Pixels

Techniques for eliminating groups of pixels using ARC/INFO functions REGIONGROUP and NIBBLE, ARC/INFO GRID, and the ArcView GIS Spatial Analyst extension are discussed in the following sections.

REGIONGROUP and NIBBLE

You can eliminate groups of pixels by using the REGIONGROUP and NIBBLE functions. REGIONGROUP gathers contiguous pixels of the same class and retains their original value in a LINK attribute. For example, the following classified grid is grouped using the REGION-GROUP function.

CLASSIFIED GRID

3	3	2	1	3	3
3	2	2	1	3	3
2	2	1	3	3	3
1	1	3	3	3	3
3	3	3	3	3	3
3	3	3	3	1	3

REGIONGROUP
$\longrightarrow$

OUTPUT GRID

1	1	2	3	4	4
1	2	2	3	4	4
2	2	3	4	4	4
3	3	4	4	4	4
4	4	4	4	4	4
4	4	4	4	5	4

VALUE	COUNT
1	6
2	4
3	25

VALUE	COUNT	LINK
1	3	3
2	5	2
3	5	1
4	22	3
5	1	1

REGIONGROUP function used to group contiguous same-value pixels.

From the above value attribute table of the output grid, group 1 has 3 contiguous pixels and was originally class 3, group 2 has 5 contiguous pixels and was originally class 2, group 3 has 5 contiguous pixels and was originally class 1, group 4 has 21 contiguous pixels and was originally class 3, and group 5 has 1 pixel that was originally class 1.

The NIBBLE function identifies small pixel groups and replaces them with a mask grid of NODATA cells. Each NODATA cell is then assigned the value from the nearest neighboring cell.

3	3	2	2	3	3
3	2	2	1	3	3
2	2	2	3	3	3
1	1	3	3	3	3
3	3	3	3	3	3
3	3	3	3	3	3

NIBBLE

2	2	2	2	3	3
2	2	2	2	3	3
2	2	2	3	3	3
2	2	3	3	3	3
3	3	3	3	3	3
3	3	3	3	3	3

		1	1	1	1
	1	1		1	1
1	1	1	1	1	1
		1	1	1	1
1	1	1	1	1	1
1	1	1	1	1	1

 NODATA CELL

Elimination of small pixel groups using the NIBBLE function.

In the following exercises using ARC/INFO GRID and the ArcView Spatial Analyst extension, REGIONGROUP is used to group contiguous same-class pixels. A NODATA mask is developed for all groups that are smaller than one hectare. The NIBBLE function is then used to fill in the small pixel groups.

GRID

1. Use the grid you named *smoothed* in the previous exercise to develop a grid of grouped same-class pixels. When the REGIONGROUP function sees the keyword FOUR, it considers for grouping only the pixels above, below, left,

and right of the candidate pixel. The keyword EIGHT tells REGIONGROUP to consider all eight surrounding pixels for grouping.

```
/***Eight directions for grouping.

Grid: groups = regiongroup(smoothed, #, EIGHT)
```

2. Next, assign NODATA to cells from all groups less than 1 ha in area. The grid cell size is 25m by 25m, or 625m^2. There are 10,000 m^2 per hectare; therefore, 1 ha is represented by 10,000/625 = 16 cells.

```
/***NODATA to cells less than one hectare.

Grid: mask = select(groups, 'count gt 16')
```

3. Use the NIBBLE function to replace NODATA cells with the nearest cell value from the classified grid.

```
Grid: gt-1ha = nibble (smoothed, mask, DATAONLY)
```

4. View results on the canvas:

```
/***Group contiguous pixels viewing all eight contiguous pixels.

Grid: groups = regiongroup(smooth, #, EIGHT)

/***Assign NODATA to groups smaller than

/***1 ha. which correspond to 16 cells.

Grid: mask = select(groups, 'count gt 16')

/***Replace NODATA cells with nearest neighbor class.

Grid: gt-1ha = nibble(smooth, mask, DATAONLY)

/***View results on the canvas.

Grid: gridshades smooth /***Smooth classified grid.

Grid: gridshades gt-1ha /***Groups under 1 ha eliminated.
```

Spatial Analyst

The exercise in this section is similar to the previous GRID exercise with the exception that Avenue requests replace GRID functions.

1. Use the grid called *smoothed* again to develop a grid of grouped same-class pixels with the .RegionGroup request. First, begin ArcView and add the *smoothed* grid theme to the view. With this grid theme active, select Analysis | Map Calculator. In the Map Calculator dialog box, enter the following expression:

```
aGRid.RegionGroup(diagNbrs, crossClass, excluded_Value
```

The .RegionGroup request in the expression requires the following parameters:

- aGrid is the input grid that will be smoothed.
- Set the diagNbrs to True if you wish to include diagonal cells in the filter's calculations. Set the flag to False if you want only the upper, lower, left, and right cells considered in the moving window.
- Set the crossClass flag to True to allow any values except excluded value to be grouped, or set the flag to False to allow only the same class values to be grouped.
- Use the excluded_Value parameter to indicate the cell value or values you wish to set to NODATA during the grouping process.

2. Try using the RegionGroup request by inputting the following in the Map Calculator:

```
( [smoothed].RegionGroup( True, False, 0) )
```

3. Press the Evaluate button to evaluate the RegionGroup expression. A new grid theme called *Map Calculation 1* will appear in the view. Make the new theme active and then select Theme | Properties to open the Theme Properties dialog box. Input *groups* for the Theme Name and click on the OK button.

4. With the groups theme active, select Analysis | Map Query. Enter the following expression in the Map Calculator:

    ```
    ( [Groups.Count] >= 16.AsGrid )
    ```

5. Double-click the Evaluate button. A value of 1 will be assigned to pixels from groups that are at least 1 ha in area, and a value of zero to pixels from groups under 1 ha.

6. Assign NODATA to the zero value pixels. With the *Map Query 1* theme active, select Analysis | Map Calculator. To change zeros to NODATA, enter the following expression in the Map Calculator:

    ```
    ( [Map Query 1]=0.AsGrid).SetNull( [Map Query 1] )
    ```

7. Double-click the Evaluate button. A new theme named *Map Calculation 2* is created. With the latter theme active, select Theme | Properties and name the theme *mask*.

8. Use the Nibble request to replace the NODATA cells with the nearest classified grid cells. Select Analysis | Map Calculator, and enter the following expression in the Map Calculator:

    ```
    ( [Smoothed].Nibble([Mask], True) )
    ```

9. Double-click the Evaluate button. A new theme name will be added to the view. This theme will represent the classified grid where all grouped pixels less than 1 ha have been eliminated.

System Integration

This chapter covers automatic batch jobs and accessing ARC/INFO programs from the ArcView menu interface. Automatic batch processing is used in the UNIX environment to rectify a scanned image and to perform automated image classification. The ARC/INFO GRID FILL command is then run from ArcView to fill in pits in an elevation grid as a precursor to hydrological analysis. The following files from the companion CD are used in the exercises and as references in this chapter.

scanquad.bil	Scanned map (also used in Chapter 6).
scanquad-world	File to be renamed to scanquad.blw world file.
scanquad.hdr	Header file describing Fairbanks, Alaska, D2 quad image.
scanquad.stx	Statistics file for Fairbanks, Alaska, D2 quad image.
fillsink.txt	ArcView Avenue script for filling with ARC/INFO GRID FILL.
emidalat	Grid of elevation estimates from Emida, Idaho.
tmrect3	Landsat Thematic Mapper band 3 grid.
tmrect4	Landsat Thematic Mapper band 4 grid.
tmrect5	Landsat Thematic Mapper band 5 grid.
tm345	Stack of Landsat Thematic Mapper grids.
batch.doc	ARC/INFO documentation file.

Automatic Batch Processing

Rectification or classification of large images often requires considerable computing time. Try running such tasks automatically whenever demands on computer resources are at low tide (e.g., evenings or weekends).

ARC/INFO Batch Jobs

Running an ARC/INFO job as a "night batch job" involves the following steps.

- Create a file of ARC batch commands to be executed.
- Change the permission of your batch file so that it is executable.
- Submit the batch job to be run at a later time.

The examples discussed in this section assume that you are working in the UNIX operating system and normally use the C-shell.

First, use a text editor (*vi* in the example) to create a batch file. The first line of the batch file should specify the UNIX shell you typically use (i.e., C-, korn, or bourne). The second line should source the file that tells UNIX about your ARC/INFO environment, such as the locations of *$ARCHOME* and the ARC/INFO *license.dat*. In the example, the user's *.cshrc* file contains these environment statements.

After the first and second line have been processed, the batch processor will know how to start ARC/INFO. The remainder of the file should contain the ARC/INFO commands you wish to execute. The example is based on *scanquad.bil*, the scanned map rectified in Chapter 6. Try rectifying the same image using batch processing.

```
vi rectify.batch

#!/bin/csh -f

source /home/user/.cshrc

/***Nearest neighbor resampling.
```

```
arc rectify scanquad.bil batch-UTM.bil nearest
```

> ➤ **NOTE:** *If you did not complete the exercise to develop the affine transformation for the* scanquad.bil *image in Chapter 6, use the UNIX mv command to move a world file containing transformation models as follows:* mv scanquad-world scanquad.blw.

In order for the batch processor to execute the commands in the *rectify.batch* file, the file must have permission set for execute. Use the UNIX *chmod* command to set the permission for execute. The syntax is *chmod u + x filename*, where *u* and *x* indicate user and add execute permission, respectively.

```
chmod u+x rectify.batch
```

Use the UNIX *at -f* command to start executing the batch file at a particular time on a particular date.

```
at -f rectify.batch 11:55pm July 3
```

The results of executing the batch file will automatically be emailed to you by the batch processor.

ARC/INFO GRID Batch Jobs

Because the grid processor expects AML command files, you must run GRID commands as an ARC macro program. For example, if you wish to run MLCLASSIFY on the image classified in Chapter 8 (*tm345*), you must first create a macro program to run the GRID commands.

Use a text editor to create an ARC macro program containing GRID commands and the batch file.

```
vi classify.aml
```

```
/***Macro to classify the stack tm345 using default options with the
mlclassify function.
```

```
classes = mlclassify(tm345,train.gsg)
```

```
&return

vi classify.batch

#!/bin/csh -f

source /home/user/.cshrc

arc grid classify.aml
```

In order for the batch processor to execute the commands in the *rectify.batch* file, the file must have permission set for execute. Use the UNIX *chmod* command to accomplish this. The syntax is *chmod u + x filename*, where *u* and *x* indicate user and add execute permission, respectively.

```
chmod u+x classify.batch
```

Use the UNIX *at -f* command to start executing the batch file at a particular time on a particular date.

```
at -f classify.batch 11:55pm December 24
```

Communication Between ArcView and ARC/INFO

Prior to ArcView GIS 3.0, the relationship between ArcView and ARC/INFO was fairly straightforward: ArcView was the user interface to ARC/INFO themes. In that context, you would use ARC/INFO to prepare and analyze map coverages and ArcView for data display and query. However, at present you may have to consider the two as separate packages. This section is focused on how ArcView communicates with ARC/INFO and accesses ARC/INFO functionality.

Assume that you wish to fill sinks (depressions or pits) in an elevation grid with the FILL command in ARC/INFO GRID prior to hydrological analysis. There are two basic methods for this task. The first is to issue an operating system command from an Avenue script. For example, you can run a script with the following two lines:

```
' Run the FILL command in ARC/INFO GRID.

system.Execute "arc \&run fill.aml"
```

The *fill.aml* macro contains the statements listed below.

```
grid /***Run the GRID module.

fill emidalat emidafill /***Fill sinks in emidalat to create emidafill.

q /***Exit from GRID.

q /***Exit from ARC/INFO.
```

Macro execution is complete when the hourglass icon disappears and the cursor returns to the ArcView window. At that point, you can use the output grid, *emidafill*, for viewing and further processing in Arc-View. If you wish, you can link the script to a new button in ArcView so that you can run the script by simply clicking the button.

In summary, the first method of filling sinks in the elevation grid treats ArcView and ARC/INFO as two separate software packages, and there is no passing of variable values between the packages.

The second way to fill sinks in an elevation grid with the FILL command is to use the Remote Procedure Call (RPC), a client/server communication system in the UNIX environment. A client process makes an RPC request and waits for a response from the server process.

In the following application, ArcView functions as the RPC client and ARC/INFO as the RPC server. The Avenue script for the application, called *fillsink.ave*, consists of four elements. First, it establishes ArcView as a client to the ARC/INFO server. This is done through a connection file and the RPClient Make request. Second, the script prepares the statement to be used by ARC/INFO GRID for executing the FILL command and uses the RPCClient Execute request to send the statement to the ARC/INFO server. Third, the script closes the connection between the ArcView client and the ARC/INFO server. Finally, the script prepares the filled grid as a new theme and draws the theme.

1. Start ArcView as a background job.

   ```
   arcview &
   ```

2. Start ARC/INFO from the same window.

   ```
   arc
   ```

3. Prepare the connection file, called *confile*, by using the IACOPEN function in AML.

   ```
   Arc: [iacopen confile]
   ```

↦ **NOTE:** *The host name for the RPC server (e.g., Host_Name), program number (e.g., 4000000), and version number (e.g., 1) are written into* confile. *If you do not see the above information in* confile, *check with your system manager about the RPC.*

4. Select File | Extensions, and click on Spatial Analyst in the Extensions window.

5. Select Scripts in the ArcView project window, and click on New. Click on the Inserting the Contents of a Text File button, and select *fillsink.txt* in the Load Script window.

6. Select Script | Properties and rename the script to *fillsink*. Click on the Compiling the Script button.

7. Choose Views in the project window, and click on New. Select View | Add Theme.

8. Select Grid Data Source as the Data Source Type. Add *emidalat* to the view.

9. Choose the *fillsink* script from the Window menu. Click on the Running the Compiled Script button. You will be asked to name the output grid for the FILL command (e.g., *emidafill*), and to verify the FILL statement. The completion of the output grid is reported in a message box. The *fillsink.ave* file is reproduced below.

```
' fillsink.txt

'------------------------------------------------------------------

' The first task is to create an ArcView client that can talk to

' the existing ARC/INFO server. The information used to establish the

' connection is returned by the ARC/INFO [IACOPEN] AML function.

' Read the connection info from the connection file, which resides in

' your current workspace. The path to the connection file is specified

' in the next line.

'------------------------------------------------------------------

iacFN = "/home/data/confile".AsFileName

iacFile = LineFile.Make (iacFN, #File_Perm_Read)

' Test again for success in getting the file.

if (iacFile = Nil) then

MsgBox.Info("Nil File object found. Can't proceed--.... Bailing
out!","")

exit

end

iacList = List.Make

iacFile.Read(iacList, 1)

iacInfo = iacList.Get(0).AsString
```

```
' Close the file.

iacFile.Close

' Extract the information from the file.

iacNode = iacInfo.Extract(0)

iacServer = ("0x" + iacInfo.Extract(1)).AsNumber

iacVersion = iacInfo.Extract(2).AsNumber

' -----------------------------------------------------------

' Establish ArcView as a client to the ARC/INFO server.

_client = RPCClient.Make (iacNode,iacServer,iacVersion)

if (_client.HasError) then

MsgBox.Info(_client.GetErrorMsg, "RPC Client Error")

exit

end

if( _client.GetTimeout = 25 ) then

_client.SetTimeout (25)

end

' -----------------------------------------------------------

' Get the output grid name.

outGrid = FileDialog.Put("outgrid".AsFileName, "*",

"What is the FILL Output Grid?")

if (outGrid = Nil) then

exit

end

theView = av.GetActiveDoc

theTheme = theView.GetActiveThemes.Get(0)
```

```
inGrid = theTheme.GetGrid

_client.Execute(1, "grid", string)

' ------------------------------------------------------------

' The following statement builds the FILL command.

aiCommand =

"FILL" ++

inGrid.GetSrcName.GetFileName.GetFullName ++

outGrid.GetFullName

MsgBox.Report(aiCommand, "GRID Command String")

' ------------------------------------------------------------

' This statement sends the GRID command to the waiting ARC/INFO server.

' The server returns control immediately to the client (this script), then

' starts to execute the command. The client can track the progress

' of the command using the string which will be returned from the server.

' This string is called the JobID.

'

' In more detail, triggering procedure number 1 on the ARC/INFO server

' starts a new job and returns its JobID. Triggering procedure number 2

' queries the ARC/INFO server for the status of an existing job specifed

' by a previously returned JobID.

JobID = _client.Execute(1, aiCommand, String)

'

' The FILL command has now started. Wait until it is finished. You can

' determine whether it is complete by triggering procedure number 2 and

' examining the return value. If that value is the string "DONE", then
```

```
' the command has finished.
'

busyWaiting = true

while (busyWaiting)

if (_client.Execute(2, JobID, String).Extract(0) = "DONE") then

busyWaiting = false

else

System.Execute("sleep 5")

end

end

' ---------------------------------------------------------------

' Close the Iac connection in ARC/INFO and the client in ArcView.

_client.Execute(1, "quit; &ty [iacclose]; quit", string)

_client.Close

'----------------------------------------------------------------

' The FILL command is finished. Prepare the output grid as a new theme.

srcString = outGrid.GetFullName

theSrcName = grid.MakeSrcName(srcString)

newGrid = Grid.Make(theSrcName)

newGTheme = GTheme.Make(newGrid)

' Add the new theme to the view and draw it.

theView.AddTheme(newGTheme)

newGTheme.SetActive(true)

newGTheme.SetVisible(true)
```

Index

Symbols

(ARCPLOT)
  color assignments in 27
.bil
  defined 99
.bil image file 25
.bip
  defined 99
.blw, .bil world file 158
.bsq
  defined 99
.cmf, color map file 28
.rmt, remap table file 27
.stx
  See image statistics file 19

Numerics

3D perspective views 57
  parameters of 58
3D surfaces 57
  lattice 63
3D views
  changing parameters 68
  creating 63
  hypsometric coloring in 81

A

accuracy assessment
  assigning reference data values 222, 224
  commission error 229
  correcting classifications 260
  creating error matrix 228
  developing reference data 216
  error table 223, 227
  omission error 229
  reference data collection methods 216
  sampling reference data, problems with 218
Add New Theme menu 63
Add Theme menu 25
Advanced Camera Orientation menu 68
  Preferences 70
affine transformation 157–158
  image adjustment 161
  of satellite image 166
  of scanned image 171
  rotation adjustment 163
  scale adjustment 161
  translation adjustment 162
affine transformation equations 158
  in rotation adjustment 163
  in scale adjustment 161
  in translation adjustment 162
affine transformation models 159, 174
  world file 158
Albers conic equal-area projection 140, 146, 152
  parameters 146
altitude
  changing 67
  in shaded relief display 82
AML command files 281
Analysis Properties dialog 238
ARC Editing environment
  defining 110
Arc Environment Properties menu 110
arc snapping
  disabling 110
ARC/INFO
  REGISTER program 164

ARC/INFO GRID extension 26
ARCEDIT
  autotracing with 110
ARCPLOT comment sign (#) 27
ARCSCAN tools 247
ARCSCAN utilities
  availability 260
ArcView-ARC/INFO
  communication exercise 282, 284
assigning colors
  custom selection 28
automatic batch jobs 279
  creating ARC macro program 281
  running GRID commands 281
  text editor commands for 280
automatic batch jobs 280
Avenue Combine request 227
Avenue scripts 181
  rectifyscript.txt 183
azimuth
  changing 67
  defined 58
  in shaded relief display 82

B

bilinear interpolation resampling 241
binary scanning
  advantages 95
  image storage 95
  threshold values in 95
  vs. grayscale scanning 95–96
BOX option 179
building value attributes (ArcView) 34

C

Camera Orientation dialog 67–68
  See also Advanced Camera Orientation
    dialog 19
Camera Preferences dialog 70
chmod command 281
Clarke 1866 spheroid 147
Classification dialog 31–32, 191
classification strategies
  supervised 187
  unsupervised 187
Clean Polygon Topology menu 106
clear-cut digitizing 130
clipping grids 236–237
CMY color model 36, 40

CMYK color model 36, 41
color images
  RGB color model 42
color infrared photography 133
  film response chart 133
color map file (.cmf) 28
color models 36, 44
color ramps 31
color translation 41
  CMYK to RGB 41
  RGB to CMY 41
COLORCHART command 40
COLORNAMES.SHD 29, 42
COMBINE function 223, 228
compart44.tif 51
compart55.tif 51
connection file 283
continuous grid, defined 30
Continuous Legend Editor 71
continuous surface
  defined 71
contrast enhancement 22
  histogram equalization 24, 25
  linear stretch 22–23, 25, 43
Convert To Grid 26
CONVERTIMAGE command 48
converting images to grids 26
coordinate systems 139, 148
  geographic 149
  in ARC/INFO 151
  in ArcView 153
  predefined (ArcView) 153
  State Plane Coordinate (SPC) system 145–146,
    149–150
  Universal Polar Stereographic (UPS) 149–150
  Universal Transverse Mercator (UTM) 145, 149
COPY command 234
copying grids 234–235
cover types 218
cubic convolution resampling 241
cutting grids
  with polygon 240–241

D

datums 147
deleting grids 234–235
DEMLATTICE command 63
dendrogram 198
DESCRIBE command 234
describing grids 234–235

digital image 19
  as grid 19
  defined 19
  pixels in 19
digital orthophotos
  color (multi-band) images 132
  defined 128
  displaying 128–130
  draping 129–130
  grayscale values 128
  linking 51
  simulating color photographs 132, 134–135
  spectral regions 132
Direct Camera/Target Input dialog 69
Discrete Legend Editor dialog box 71
discrete surface
  defined 71
displaying grids 26, 30
displaying images 24
dpi
  file size in 97
  vs. image storage space 97
draping features
  See 3D draping 19
drawing commands
  with SURFACEDRAPE 80

E

edge enhancement 264, 266
  range filter 264
Edit Arcs & Nodes menu 105, 110
ellipsoid 147
error matrix 215, 218
  consumer's accuracy 230
  generating 218
  producer's accuracy 229
  stratified random sampling 218
Export command 48

F

FILL command 279, 282–283, 285
filtering grids 263
  See also edge enhancement 19
  See also smoothing filters 19
floating point grid
  See continuous grid 30
FOCALRANGE function 267
FocalStats request 267
  parameters 268

Fuzzy Tolerance 106

G

generalization
  defined 107
GENERATE command 173
generating an error matrix 218
grayscale display 26
grayscale images
  defined 21
  See also single-band images 19
grid cells
  defined 19
grid display
  See displaying grids 19
grid display commands 26
grid editing
  changing straightening properties 106
grid management 233–237, 240–243
  resampling 244
grid management 242
Grid Property parameters 71
Grid Property Sheet 71
Grid Property Sheet dialog 65, 68
grid resampling
  See resampling grids 19
GRIDCOMPOSITE command 29
  defined 26
GRIDPAINT command 26–27
GRIDQUERY command 26, 29, 199
GRIDSHADES command 26, 28
ground control points (GCP) 159, 164
GRS 80 (Geodetic Reference System 1980)
  spheroid 148

H

heads-up digitizing
  advantages 101
  vs. tablet digitizing 101
hierarchical cluster analysis 196
Hillshade
  altering symbology 88
HILLSHADE command 63
Hillshade function 86
histogram equalization 24
HLS color model 36, 39
Hot Link tool 50
  linking digital orthophotos 51
  linking objects 50

HSV color model 29, 36, 38, 44
hypsometric coloring 81

I

IACOPEN function (AML) 284
identity display 22
Identity tool 183
image classification
   accuracy assessment 215
IMAGE command 24, 42–43, 134
image enhancement 22
image rectification 139
image statistics file 22, 25, 43
   contents of 25
Image/Coverage windows 165
IMAGEGRID command 26
IMAGEVIEW command 134
ISOCLUSTER function 196

K

KILL command 234

L

Lambert conic conformal projection 140, 146, 154
   parameters 146
Landsat Thematic Mapper image
   spectral bands/regions chart 133
Legend Editor 23, 26, 30
Line Properties dialog 108
   Dash parameter 109
   Gap parameter 109
   Hole parameter 109
   Value parameter 109
   Variance parameter 109
   Width parameter 109
line straightening
   defined 107
linear stretch 22–23, 25–26
   default 47
LINECOLOR command 37
Link Actions window 171

M

majority filter 268, 270
MajorityFilter request 272
   parameters 272
Map Calculator (ArcView) 35
map projections 139

Albers conic equal-area 146
aspects 140–141
aspects, equatorial 141
aspects, oblique 141
aspects, polar 141
azimuthal 140
central meridian 143, 145
central parallel 143
classes 140
conformal 140, 143
conic 140–141
cylindrical 140
defined 139
degree of distortion in 142
equal area 140
equidistant 140
in ARC/INFO 151
in ArcView 153
Lambert conic conformal 146
line(s) of tangency 140–141
Mercator 143
polyconic 147
predefined 153
principal scale 140
secant conic projection 140
simple conic projection 140
standard meridian 141, 143
standard parallel(s) 141, 143
transverse Mercator 145
Map Tools menu 63
MAPPROJECTION command 153
MARKERCOLOR command 37
maximum likelihood rule 208, 210
   statistical formula 208
Mercator map projection 143
   parameters 143
MERGE command 241
MLCLASSIFY function 281
multi-band images 132
multispectral
   defined 194

N

NAD 27 (North American Datum of 1927)
   spheroid 148
national map accuracy standards 151
nearest neighbor resampling 241–242
   advantages 242
NEATLINE command 152
NEATLINEGRID command 152

NIBBLE function 275
    defined 274
Nibble request 278
NODATA 277
NODATA mask 274–275
node snapping
    disabling 110
noise regions
    defined 252
    defining 253
    examples 252
    removing 253–254, 256
North American Datum of 1927 150
North American Vertical Datum of 1988 (NAVD 88) spheroid 148

O

Overlay window (REGISTER) 165

P

packaging scanned images 99
panchromatic display 24–26
phantom graphics 252–253
pixel values 22–23, 206
    assigning colors 42
    calculating limits 23
    defined 21
    limits 23
    mapping 22
pixels
    correcting classification 261
    defined 19
polyconic map projection
    parameters 147
pound sign (#) 199, 211
PROJECTION command 151
Projection Properties dialog 155

Q

Query Builder 35, 54

R

random sampling
    aerial photography exercise 218
    assigning truth classes 222
    converting points to grid 223
raster line

changing parameters of 107
    defined 107
    determining width 107, 109
raster to vector data conversion 102
RECLASS function 201, 211
RECTIFY command 179
rectifying grids 181
rectifying images 176
    resampling, See resampling images 176
rectifying images 179–180
Reduce Noise menu 253
reference data
    aerial photography 218
    random sampling 216
    vector GIS polygons 217
REGIONGROUP function 273, 275
    defined 273
Regions tool 254
REGISTER command 164, 174
Registration Results window 169
remap table 27–28
Remote Procedure Call (RPC) 283
remotely sensed images
    advantages 113
    displaying orthophotos 128
    See digital orthophotos
    resolution types 121
RENAME command 234
renaming grids 234–235
RESAMPLE command
    bilinear method 242
    cubic method 242
    nearest 242
RESAMPLE command 242
resampling grids
    bilinear interpolation 241
    cubic convolution 241
    defined 241
    nearest neighbor 241
resampling images
    bilinear interpolation 176–177
    cubic convolution 176, 178
    nearest neighbor 176–177
    options 176
resemblance matrices
    computing spectral distance 197
resolution
    radiometric, explained 124
    spatial 121
    spectral 121, 123

temporal 121
temporal, explained 124
RGB color model 36–37, 44–45
RGB values 36
    in .cmf files 28
RMS error 170
RPC client 283
RPC server 283
RPCClient Execute request 283
RPCClient Make request 283

S

satellite images
    planned high spatial resolution imaging
        systems 125
    quality control check 126
    types chart 125
saving transformation models 171
scanned images
    conversion to ARC/INFO grid 102
    editing 102
    generalization tolerance 107
    line 107
    tracing 102, 104, 110
    tracing arcs 102
scanning
    computing file sizes 99
    image preparation 248
    image storage 93
    image storage, binary 94–95
    image storage, bits and bytes explained 94
    image storage, color 94
    image storage, grayscale 94
    packaging images, generic formats 99
    packaging images, proprietary formats 100
    raster to vector data conversion 102
screen values 24
SCREENSAVE command 48
SELECTPOLYGON tool 240
SETWINDOW command 236
SHADECOLOR command 41
shaded relief 82, 85–86
    contrast enhancements 85
    variations 88
shaded relief images 57
SHOW command 41
single-band images
    displaying 128
Sketch menu 255

drawing tools 255
filling tools 255
slide shows
    creating 48
    displaying 50
smoothing filters
    applications 269
Source Type
    selecting 25
Spatial Analyst extension 26, 30, 181, 187, 204,
        227, 233, 266, 277
    displaying shaded relief 86
    shaded relief images 57
Specify Color menu 45
speckles 248, 252
    defining 252
    distinguishing 250
    removing 248, 250–252, 256
spectral classes 188, 196, 203
    assigning cover types 191, 199
    hierarchical cluster analysis 196
    merging 198
    pixel values of 189
    resemblance matrices 197
    resemblance matrices, dendrogram 198
    single vs. multiband classification 194
spectral clustering 189, 196
spectral distance 199
    calculating 197
    dendrogram 198
spectral reflectance
    defined 115
    in spectral regions 115
    surface reflectance response chart 120
spectral regions
    chart 116
    comparisons 116
    defined 114
spheroid 147
SPHEROID command 151
Spot XS image
    spectral bands/regions chart 134
State Plane Coordinate (SPC) system 145
stitching grids 236
    with polygon 241
Straighten Properties menu 106
supervised classification 203
    defined 187
    delineating training areas 205
    developing training area statistics 206

maximum likelihood rule 208
predicting pixel class 211
selecting training areas 204
use of training areas 203
Surface Properties menu 64
surface z-scale 60
defined 60
SURFACEDRAPE command 80
SURFACEOBSERVER command 80
SURFACEZSCALE command 80

T

TEXTCOLOR command 37
Theme Manager menu 63
Tracing Environment menu 106
training area statistics
developing 206
maximum likelihood rule 211
training areas
defined 203
delineation 205
pixel selection 204
TRANSFORM command 158
transverse Mercator map projection 145
parameters 145

U

Universal Transverse Mercator (UTM)
projection 179
UNIX operating system 280
unsupervised classification 188, 194, 203
assigning cover types 187, 191
defined 187
multispectral 196
spectral classes 187
spectral clustering 187, 189
UTM coordinate system 188
UTM projection 157

V

value attribute table (.vat) 34
View Properties dialog 155
viewing angle
defined 59

W

WGS (World Geodetic System) 84 spheroid 147,
148

More OnWord Press Titles

Computing/Business

Lotus Notes for Web Workgroups
$34.95

Mapping with Microsoft Office
$29.95 Includes Disk

*The Tightwad's Guide to Free Email
and Other Cool Internet Stuff*
$19.95

Geographic Information Systems (GIS)

GIS: A Visual Approach
$39.95

The GIS Book, 4E
$39.95

INSIDE MapInfo Professional
$49.95 Includes CD-ROM

MapBasic Developer's Guide
$49.95 Includes Disk

*Raster Imagery in Geographic Information
Systems* Includes color inserts
$59.95

INSIDE ArcView GIS, 2E
$44.95 Includes CD-ROM

ArcView GIS Exercise Book, 2E
$49.95 Includes CD-ROM

ArcView GIS/Avenue Developer's Guide, 2E
$49.95

*ArcView GIS/Avenue Programmer's
Reference, 2E*
$49.95

ArcView GIS /Avenue Scripts: The Disk, 2E
Disk $99.00

ARC/INFO Quick Reference
$24.95

INSIDE ARC/INFO
$59.95 Includes CD-ROM

*Exploring Spatial Analysis in Geographic
Information Systems*
$49.95

*Processing Digital Images in GIS:
A Tutorial for ArcView and ARC/INFO*
$49.95

*Cartographic Design Using ArcView GIS and
ARC/INFO: Making Better Maps*
$49.95

MicroStation

INSIDE MicroStation 95, 4E
$39.95 Includes Disk

MicroStation 95 Exercise Book
$39.95 Includes Disk
Optional Instructor's Guide $14.95

MicroStation 95 Quick Reference
$24.95

MicroStation 95 Productivity Book
$49.95

Adventures in MicroStation 3D
$49.95 Includes CD-ROM

MicroStation for AutoCAD Users, 2E
$34.95

MicroStation Exercise Book 5.X
$34.95 Includes Disk
Optional Instructor's Guide $14.95

MicroStation Reference Guide 5.X
$18.95

Build Cell for 5.X
Software $69.95

101 MDL Commands (5.X and 95)
Executable Disk $101.00
Source Disks (6) $259.95

Pro/ENGINEER and Pro/JR.

*Automating Design in Pro/ENGINEER
with Pro/PROGRAM*
$59.95 Includes CD-ROM

INSIDE Pro/ENGINEER, 3E
$49.95 Includes Disk

Pro/ENGINEER Exercise Book, 2E
$39.95 Includes Disk

Pro/ENGINEER Quick Reference, 2E
$24.95

Thinking Pro/ENGINEER
$49.95

Pro/ENGINEER Tips and Techniques
$59.95

INSIDE Pro/JR.
$49.95

*INSIDE Pro/SURFACE: Moving from Solid
Modeling to Surface Design*
$90.00

FEA Made Easy with Pro/MECHANICA
$90.00

Softdesk

INSIDE Softdesk Architectural
$49.95 Includes Disk

Softdesk Architecture 1 Certified Courseware
$34.95 Includes CD-ROM

Softdesk Architecture 2 Certified Courseware
$34.95 Includes CD-ROM

INSIDE Softdesk Civil
$49.95 Includes Disk

Softdesk Civil 1 Certified Courseware
$34.95 Includes CD-ROM

Softdesk Civil 2 Certified Courseware
$34.95 Includes CD-ROM

Other CAD

Manager's Guide to Computer-Aided
Engineering
$49.95

Fallingwater in 3D Studio
$39.95 Includes Disk

INSIDE TriSpectives Technical
$49.95

Interleaf

INSIDE Interleaf (v. 6)
$49.95 Includes Disk

Interleaf Quick Reference (v. 6)
$24.95

Interleaf Exercise Book (v. 5)
$39.95 Includes Disk

Interleaf Tips and Tricks (v. 5)
$49.95 Includes Disk

Adventurer's Guide to Interleaf LISP
$49.95 Includes Disk

Windows NT

Windows NT for the Technical Professional
$39.95

SunSoft Solaris

SunSoft Solaris 2.* for Managers and
Administrators
$34.95

SunSoft Solaris 2.* User's Guide
$29.95 Includes Disk

SunSoft Solaris 2.* Quick Reference
$18.95

Five Steps to SunSoft Solaris 2.*
$24.95 Includes Disk

SunSoft Solaris 2.* for Windows Users
$24.95

HP-UX

HP-UX User's Guide
$29.95

Five Steps to HP-UX
$24.95 Includes Disk

OnWord Press Distribution

End Users/User Groups/Corporate Sales

OnWord Press books are available worldwide to end users, user groups, and corporate accounts from local booksellers or from SoftStore/CADNEWS Bookstore. Call 1-800-CADNEWS (1-800-223-6397) or 505-474-5120; fax 505-474-5020; write to SoftStore, Inc., 2530 Camino Entrada, Santa Fe, New Mexico 87505-4835, USA or e-mail orders@hmp.com. SoftStore, Inc., is a High Mountain Press Company.

Wholesale, Including Overseas Distribution

High Mountain Press distributes OnWord Press books internationally. For terms call 1-800-4-ONWORD (1-800-466-9673) or 505-474-5130; fax to 505-474-5030; e-mail to orders@hmp.com; or write to High Mountain Press, 2530 Camino Entrada, Santa Fe, NM 87505-4835, USA.

Comments and Corrections

Your comments can help us make better products. If you find an error, or have a comment or a query for the authors, please write to us at the address below or call us at 1-800-223-6397.

On the Internet: http://www.hmp.com

OnWord Press, 2530 Camino Entrada, Santa Fe, NM 87505-4835 USA